I0032697

V. (Valentin) Magnan

On Alcoholism

The Various Forms of Alcoholic Delirium and Their Treatment

V. (Valentin) Magnan

On Alcoholism
The Various Forms of Alcoholic Delirium and Their Treatment

ISBN/EAN: 9783744670289

Printed in Europe, USA, Canada, Australia, Japan

Cover: Foto ©berggeist007 / pixelio.de

More available books at **www.hansebooks.com**

ON

ALCOHOLISM

THE

VARIOUS FORMS OF ALCOHOLIC DELIRIUM

AND

THEIR TREATMENT

BY

DR. V. MAGNAN,

PHYSICIAN TO ST. ANNE ASYLUM, PARIS; LAUREATE OF THE INSTITUTE, ETC.

TRANSLATED BY

W. S. GREENFIELD, M.D., M.R.C.P.

LONDON:

H. K. LEWIS, 136 GOWER STREET, W.C.

1876.

TRANSLATOR'S PREFACE.

In bringing the following work before the notice of the profession in England, I have been influenced in part by the desire to fill up an acknowledged gap in medical literature. There are few medical subjects on which less has been written than on delirium tremens and chronic alcoholism, and that in spite of their constant presence, and their great social importance. It is true that during the last two or three years considerable attention has been devoted to the subject of the uses of alcohol in health and disease, and to the evils resulting from excessive drinking. But, as yet, this increased attention has not resulted in any notable addition to our medical literature.

The systematic English works on the subject are scanty and meagre, and two years ago the same might with truth have been said of the papers and reported cases to be found in periodical literature. Possibly this fact may be due to the limited view which has been taken of the subject, and the want of more careful clinical observation specially directed to it. The extreme frequency, one might almost say the constancy, of the alcoholic element in disease, combined with the frequency of drunkenness and delirium tremens, have also perhaps led to some of this neglect. The effects of continued drinking on the viscera have been regarded as almost limited to the production of cirrhosis of the liver, chronic gastritis, and granular kidney ; *delirium tremens* has been considered a disease of so well marked and uniform character as to require no special study. Nevertheless, the views as to its treatment, and the statements as to its clinical phenomena will be found to be widely at variance in different hospitals, and amongst different schools of medicine. Doubtless there have been many who have taken a wider view of the subject, and have arrived at a more intimate knowledge of the action of the poison in its less common forms, but with few exceptions they have not placed their observations on record. Thus alcoholic paraplegia was recorded and described some years ago by Dr. Wilks,* and other nervous disorders by Dr. Handfield Jones, but these affections have hardly been noticed in text-books of medicine.†

The systematic works which treat especially of the effects of alcohol, are Dr. Anstie's book on *Stimulants and Narcotics*‡ and Dr. Marcet's *On Chronic Alcoholic Intoxication*.§ The former, however, deals mainly with the physiological action of alcohol, and the latter must now be considered as somewhat meagre. Some valuable clinical remarks will be found in Dr. Handfield Jones's work on *Functional Nervous Disorders;* and articles, for the most part brief, in the various text books of Medicine and Surgery may also be referred to.

The widespread interest which has been manifested during the past two or three years in the various questions connected with the use and abuse of alcohol has shown that the subject is one on which the public in general as well as the profession feels strongly. Yet it is in great measure to the medical profession that this increased interest is due. The declaration as to the evil effects of the indiscriminate administration of alcohol as a medicine, which was signed by a large number of leading practitioners, although regarded by many as ill-advised, undoubtedly did much to draw popular attention to the subject, and this has been kept alive by articles in newspapers, magazines, and reviews which have largely echoed the opinions of the profession.

It may be worth while to mention some of the various ques-

---

* *Lancet* 1872. Vol. I., p. 320.

† It is true that this form of disease (alcoholic paraplegia) appears to be of far less frequent occurrence in England than in some other countries. I am informed by a friend, a physician of great experience in Sydney, New South Wales, that it is not at all uncommon in Australia, and is there a well recognized form of disease. Experiments on dogs and other animals have shewn that paraplegia is one of the more immediate physiological effects of alcohol upon them. Hence it is not surprising that its long continued use in man should result in degeneration of the cord, which in some cases may be productive of paraplegia, in others of a condition closely allied to locomotor ataxy.

‡ London, Macmillan, 1864.   § 2nd edition. London, 1862.

tions to which the attention of the community, but still more of the profession has been directed. These may be classified as they relate to the *use* or *abuse* of alcohol. Under the former we may range the inquiries into its physiological action on man and animals by means of experiment and observation, together with the question of its use as an article of diet; the investigations as to its employment in disease, especially fever, and its power of lowering the temperature, increasing the force of the pulse, and preventing or controlling the waste of tissue.

As regards its abuse, apart from the evils resulting to the individual in the form of visceral degeneration and disease, which will especially occupy our attention, there are the questions of its excessive or needless employment as a food or as a medicine : the evils of the multiplication of temptations to drinking, and the means of reducing them and of checking drunkenness. And, lastly, the legislative enactments for the punishment of drunkennesss, and for the restraint or incarceration of habitual drunkards. To these might perhaps be added the more remote questions of the effects of drinking in the general deterioration of the population, and the association of drunkenness, crime and insanity in the offspring of drunken parents.

The time has not yet arrived for the collection and summing up of all the work which has been done in these various branches of the subject, since the work is still in progress. But we may venture to mention merely some of the more important work on each subject, though only briefly to indicate it. The physiological action of alcohol on the healthy system has been the study of various investigators, amongst whom in England the names of Parkes, Anstie, Edward Smith and Brunton must always rank amongst the first. The continued and exact investigations of the lamented Dr. Parkes on the effects of alcohol on digestion, excretion and the production of force, and the action of alcohol on the complex vital functions, have formed a groundwork of knowledge of the highest value, and which must be of the greatest use to all future observers. To Dr. Anstie we not only owe much valuable original observation both physiological and clinical, but to him is due the credit of having for some time been almost alone in keeping the subject before the notice of the profession. Since his untimely death, Dr. Brunton has done good service in a similar direction, and to him we are indebted for a recent paper, read before the Medical Society of London, which well expounded our present knowledge of the physiological action of alcohol, whilst the discussion upon it shewed the general interest in the subject. It is not our purpose here to apportion to each the credit for research in one or other part of the subject, nor to take up in detail the various points which have received attention from these and other observers. But we may mention also the names of Drs. Ringer and Rickards as having made experiments on the action of alcohol in relation to the elevation and depression of temperature.

The value of the various forms of alcoholic beverages as articles of diet has been largely discussed of late, and this from a more scientific standpoint than previously. The researches of Dr. Parkes, to which we have already referred, the observations made during the Ashantee war, and the discussion on the issue of the spirit ration, have much increased our knowledge and have diffused that already acquired.

With respect to the value of alcohol as an article of diet, some members of our profession, as for example, Sir Henry Thompson, have been led by a long course of observation, to conclude that even in moderate quantity, it is not only needless but injurious in its ultimate effects upon the system. The statistics of insurance companies seem to point in the same direction, so far as the healthy portion of the population is concerned, though these, like all other statistics, are liable to misconception. The profession is greatly indebted to Dr. B. W. Richardson for bringing together and popularizing the results of his own investigations and those of others on these and various other points connected with the action of alcohol, in the recent Cantor Lectures.*

Besides the declaration of the medical profession before re-

---

* *On Alcohol:*—A course of six Cantor Lectures delivered before the Society of Arts, by Benjamin W. Richardson, M.D., F.R.S. London, 1876.

ferred to, we are bound to mention with respect the recommendation made by the College of Physicians for a Government enquiry into the evils of intemperance and the means of preventing it. The legislation of late years, although it has not gone so far as some have wished, and undoubtedly still leaves much to be desired, has been in the direction of controlling the supply of drink and imposing penalties for intoxication.

These facts show that there is a steadily increasing current of popular opinion, which will undoubtedly gain in force and lead to further legislation; and we may hope that this will take the form of diminishing the temptations to drink and the facilities for obtaining it at late hours. The immense improvement effected by the earlier hour of closing of public houses was so marked in hospital experience, that it is greatly to be regretted that the hours should again have been lengthened.

On one other point public opinion is not as yet ripe for legislation, and even in our own profession there is considerable difference of view upon it, viz. the control to be exercised by the State over habitual drunkards. Nor can we expect that until more unanimity of opinion prevails, both as to its desirability and as to the cases in which it is likely to prove beneficial, there will be any sufficient influence on public opinion to lead to legislation.* It is to be hoped that the efforts of the committee of the British Medical Association on this question, will be ultimately productive of a good result.†

Finally, the terrible results of drunkenness on the children of drunken parents, and the close relation of vice, crime and insanity in their offspring with the habits of the parents, on which Dr. Maudsley and others have given such startling facts, will tend to impress the necessity of checking the evil as far as possible.

---

* Much needless obscurity has been thrown over this question, by the confusion of such totally distinct forms of disease as dipsomania and chronic alcoholism: even medical journals have been guilty of this error, which has added much to the difficulty of a clear comprehension of the subject.

† *Brit. Med. Journal* Aug., 12, 1876, p. 214.

All these facts, pointing as they do to an increase in the interest in the subject, lead us to hope that a systematic work dealing with the subject in all its scientific bearings will at no distant
date be possible. At the present time the data are not sufficiently complete.

The scope of the present work does not of course include the
several topics of interest in relation to alcohol which have been
indicated above. It deals, in fact, with only one branch of the
subject, viz., the poisonous action of alcohol in an overdose, and
the conditions of disease resulting from it.

It may be well to mention the opportunities which have fallen
to the author for the observation and investigation of alcoholism.
Unquestionably one of the main reasons why so little attention
is given to the subject in English medical literature, is the
want of opportunity of continuous observation on any large
number of cases of alcoholism. At some time or other every
hospital physician or surgeon sees a considerable number of cases
of delirium tremens. An especially large number, however,
enter the surgical wards for some accidental or self-inflicted
injuries—or the malady breaks out under the influence of traumatic conditions. It is to the injury rather than to the delirium
that the attention of the surgeon is directed, and some symptoms
of the latter may be ascribed to the former. So too in the cases
coming under the care of the physician, those of simple alcoholic
delirium are rapidly cured, often before his visit, and the more
severe and often fatal cases of delirium tremens are either
complicated by some other disease, or appear to be so. Hence
the well marked cases of severe uncomplicated delirium tremens
coming under the notice of any individual physician are comparatively few.

What is of still greater importance is that in very few cases
do we have the opportunity of seeing the same patient in successive attacks, and of watching the progressive effect of continuous
alcoholic poisoning. We see indeed daily the effects of alcohol
on the constitution, and its various symptoms in those who indulge in it to excess; the post-mortem room reveals to us only

too often the resulting degeneration of the organs, in their several stages ; and frequently we are convinced of the presence of such degenerations during the life of our patients, although we can get no positive proof of their existence ; but very rarely can we trace for ourselves the whole course of the downward progress, or even mark its steps at definite epochs.

Hence it is that opportunities for systematic observation such as have been enjoyed by the author of this work, are of exceptional value, and it is desirable for us to give some account of them for the benefit of the English reader.

The Bureau d'Admission of the department of the Seine at the St. Anne Asylum in Paris, of which Dr. Magnan is one of the two physicians, is an institution to which no exact parallel exists in England. To it are brought all the cases of insanity previous to their admission to the various public asylums, and all cases of acute delirium and mania which fall under the care of the police in Paris. It is here that they are examined and their admission or rejection decided upon ; if admitted they are drafted to the one or other of the asylums which is the most suited to the class of the patient or the form of his malady.

The Bureau d'Admission is quite distinct from the St. Anne Asylum itself, and under altogether different administration. In order to provide accommodation for the temporary lodgment of patients on their way to other asylums, and also for the reception of the more acute cases, it is provided with about 50 beds, and is fitted up in every way as a small asylum.

Here then are brought all the cases of delirium tremens and of simple alcoholic delirium which fall under the notice of the police, and a large number from the lower and middle classes, and here they are treated until their recovery. Cases too of fever with delirium are not unfrequent, and it need scarcely be said that acute delirious mania is also often seen.

Hence it comes to pass that a very large proportion of all the cases of delirium tremens occurring in Paris and its vicinity come under observation here, and this not only in one attack, but again and again, and when at last by repeated attacks they have become mentally deranged or greatly weakened, they again come under notice for transference to asylums. The results of such opportunities of observation could scarcely fail to be productive of an increase in our knowledge, and their value is necessarily augmented by the fact of their being utilized by experienced alienists, and seen side by side with other forms of acute delirium.

It should be added that there is also an out-patient department, to which not only cases of mental derangement but of all forms of nervous disorder, especially epilepsy, are gratuitously admitted, and to this department those who have before been under treatment for the acuter symptoms of alcoholism resort for the treatment of the various nervous disorders induced by their habits.

On certain points the observations of the author will be found at variance with the opinions currently received in England. That *delirium tremens* when uncomplicated by inflammatory disease is a delirium without fever is an opinion commonly stated in books, and upon this belief not only diagnosis but treatment are not unfrequently based. Were it not so conclusively shown by the cases here recorded that this view is incorrect, it would have been easy to adduce cases which have fallen under my own observation in which elevation of temperature has been a well-marked symptom.* The importance of the recognition of this fact, or rather of the converse, that elevation of temperature in delirium tremens does not necessarily indicate visceral inflammation, might also have been illustrated by cases in which alcoholic stimulation has been resorted to on the ground of supposed pneumonia, the non-existence of which was only discovered after the fatal issue. The error has doubtless arisen from non-observation or observation limited to cases of acute alcoholic delirium, in which the elevation is slight or wanting.†

* The profuse perspiration, which causes great surface cooling, together with the agitation of the patient, often render observations taken with the thermometer in the axilla either impossible or valueless. The temperature ought in these cases to be taken in the rectum.

† It must be remembered that the *immediate* effect of alcohol poisoning is to

It may perhaps be thought that the value of the work is marred by the fact of the differences in the forms of alcoholic beverage in use in France from those most commonly taken in England, and especially by the reference to the effects of *absinthe*, which is comparatively little used in this country. That this objection is not generally considered a serious one is shown by the fact that the work has already been translated into German, Spanish, Italian and Russian. But further consideration will, I believe, show that it is no disadvantage. The broad features of alcoholic poisoning whether acute or chronic are similar whatever the form in which spirit is taken in excess. The effects, especially in the acuter forms, are undoubtedly largely dominated and modified by national and individual temperament, but however varied the accent and inflection, the language, so to speak, is the same. So too, there is no doubt that each form of spirit has to some extent a different action, and that there is a real difference in the immediate effects of beers, wines, and spirits, which is not wholly accounted for by the different degree of dilution of the alcohol contained in each.

It would be a great gain to our knowledge both as regards the employment of the various forms of spirituous liquors in medicine and diet, and our scientific knowledge of the effects of alcohol itself, if we could precisely determine the effects, physiological and toxic, of the various ingredients of the beverages most commonly employed in this country, and ascertain how far they tend to modify the action of alcohol in the system, or to produce additional symptoms. The very minute quantity of these substances, except where other liquors or " bitters" are purposely added to them, or where adulteration is practised, renders this a difficult task. Certain of these substances are known, and to some extent the variety in action of different spirits and wines are matters of popular knowledge, but the precise physiological effects of each in repeated small doses are as yet but little known.

lower the temperature. For further details, I may refer the reader to Wunderlich's *Medical Thermometry*. (New Syd. Soc. Trans.), pp. 117 and 137.

The chemical researches of Drs. Dupré and Thudichum have, however, paved the way for further physiological investigation in this subject, and its completion is we trust not far distant. For information on this subject we must refer the reader to Dr. Richardson's able lectures,* where also will be found some valuable remarks on the effects of absinthe.

But so far from its being a disadvantage that the substance which is so commonly taken in France in combination with alcohol is especially discussed in this essay, it is we think, a gain, as it affords a clear and definite example of the specific effects of one form of adulteration, and leads us to hope that a clearer light may be likewise thrown on the action of other analogous substances. It is very probable that the various forms of "bitters" and the like which are so largely consumed in this country have some action similar to that of absinthe, although less potent and deadly, nor must it be supposed that the use of the latter itself is either unknown or very rare in this country.†

I have thought it desirable to say thus much in explanation of the scope and objects of the work, and I can but trust that the many defects in the translation of which I am fully conscious, will not be such as to interfere with the real value of the work.

If in any way the work tends to the diffusion of knowledge of the deadly effects of alcohol, and to an increased attention to the results of excess in drinking and a greater activity in the efforts now being made to arrest its progress; and if a perusal of the cases herein recorded carries the inevitable conviction that means other than moral should be employed to restrain the chronic drunkard, the work will not be in vain, even apart from its more immediate scientific object.

* Richardson *loc. cit.* Lect. V. pp. 76 *et seq.*
† Richardson *loc. cit.*, p. 78.

CONTENTS.

PAGE

TRANSLATOR'S PREFACE

Introduction 1

Plan and division of the work 3

CHAPTER I.

IMMEDIATE ACTION OF ALCOHOLIC BEVERAGES; DRUNKENNESS.—
PROLONGED ACTION; ALCOHOLIC DELIRIUM.

ARTICLE I.—Drunkenness in man; period of excitement; intellectual
disturbance; incoherence; paralysis; anæsthesia; coma . . . 5

Convulsive drunkenness 6

ARTICLE II.—Drunkenness and alcoholic delirium in the dog . . 7

1. Drunkenness; excitement; stupor; paralysis; anæsthesia; comatose
condition 7

2. Alcoholic delirium; changed behaviour; irritability; hallucinations;
delirium; trembling; vomiting; gastro-enteritis 8

Modes of termination; death by cold; by broncho-pneumonia; by asphyxia; by accident; by cachexia 10

Pathological anatomy; acute lesions; presence of alcohol in the organs 11

Chronic lesions, (fatty degeneration, sclerosis) 12

Experiment I.—Prolonged action of alcohol in the dog 18

ARTICLE III.—Action of absinthe on animals; muscular shocks; .
vertigo; epileptic attack; hallucinations; delirium 22

Experiment II.—Epileptic attacks and hallucinations in the intervals
between the attacks, due to injection of essence of absinthe into the
stomach 26

Experiment III.—Epileptic attacks and hallucinations under the influence of essence of absinthe injected into the veins. 26

Appearance of hallucinations in man (as in the animal), more rapid under
the influence of absinthe, than under that of alcohol 29

Case I.—Rapid appearance of delirium under the influence of absinthe
liqueur 29

CHAPTER II.

ALCOHOLIC DELIRIUM IN MAN.

Change of character; irritability; hallucinations and delirium . . 33

ARTICLE I.—General characters of alcoholic delirium . . . 33

1. Painful character of the hallucinations 33

2. Changeable character of the hallucinations 34

3. Hallucinations have for their subject either the daily occupation of
the patient, or the subjects of predominant interest at the time . . 34

Maniacal, melancholic and stupid forms 37

Secondary importance of these three forms 38

PAGE

ARTICLE II.—Mode of evolution of the sensory disturbances . . . 39
  1. Sensory disturbances examined as a whole 39
  2. Sensory disturbances examined in relation to the peculiarities which
    they offer in each of the senses 40
ARTICLE III.—Classification of patients affected with alcoholic delirium
  in three groups, according to the antecedents of the patient and the
  course and termination of the disease 45
  Group 1. Patients with alcoholic delirium with easy, complete and rapid
    convalescence . : 45
    Case II.—Simple alcoholic delirium ; cure 46
    Case III.—Two attacks of alcoholic delirium ; cure 48
  Group 2. Patients affected with alcoholic delirium with slow convales-
    cence and tendency to relapse 50
    Case IV.—Alcoholic delirium of maniacal, melancholic and stupid forms 51
    Case V.—Sixth attack of alcoholic delirium ; slow recovery . . , 55
    Case VI.—Repeated attacks of alcoholic delirium 58
  Delirium of depressive form ; suicidal and homicidal tendencies . . 61
  Delirium of expansive form ; ambitious ideas 62
  Group 3. *Predisposed* patients, affected with alcoholic delirium with fre-
    quent relapses, &c. 63
    Case VII.—Hereditary predisposition ; repeated attacks of alcoholic de-
    lirium 65
    Case VIII.—Inherited neurotic tendency ; relapses of alcoholic delirium 69
ARTICLE IV.—Action of absinthe on man ; muscular shocks ; vertigo ;
  epileptic attacks ; absinthic epilepsy bears no relation to the trembling
  or hallucination ; delirium . . . . . . : . . . 73
    Case IX. 78
    Case X. 79
    Case XI. 81
    Case XII. . . . . , 87
    Case XIII. 90
ARTICLE V.—Treatment.
  1. Of Drunkenness 92
  2. Of alcoholic delirium 93
    Agents derived from the materia medica 93
    Confinement ; occasions for its employment ; its duration . . 94
    Prophylaxis 98

CHAPTER III.

FEBRILE DELIRIUM TREMENS.

ARTICLE I.—Symptoms and diagnosis of febrile delirium tremens . . 102
  The severity of the delirium in febrile delirium tremens does not neces-
    sarily correspond with its gravity 104
  The two essential characters : 1st. Fever, appreciable by means of the
    thermometer . . . : 104
  2nd. Trembling with small shocks ; tremors and fibrillar contractions of
    the muscles, especially appreciable by the touch 105
  Case XIV.—Febrile delirium tremens ; high temperature ; death ; au-
    topsy 109
    Chart of temperature and pulse, Fig. I. 113
  Case XV.—Febrile delirium tremens ; death ; autopsy 118
    Chart of temperature and pulse, Fig. II. 120

PAGE
Case XVI.—Febrile delirium tremens; high temperature; death . . 122
Chart of temperature and pulse 123
Case XVII.—Febrile delirium tremens; death, autopsy 124
Analysis of organs for alcohol 125
Case XVIII.—Febrile delirium tremens; autopsy 127
Case XIX.—Epileptic attack; febrile delirium tremens; cure; chronic alcoholism 130
Chart of temperature and pulse 133
Table shewing the comparative degrees of anæsthesia of the limbs . 136
Case XX.—Febrile delirium tremens; cure 139
Chart of temperature and pulse 140
ARTICLE II.—Differential diagnosis of febrile delirium tremens and delirium tremens associated with intercurrent affections and injuries . 141
ARTICLE III.—Treatment of febrile delirium tremens 143
1. Protection of the patient against himself, and prevention of injury to those around him 145
The camisole (strait-waistcoat); its action when combined with fixation to the bed 146
Case XXI.—Maniacal attack *without fever*; fixation to the bed; asphyxia; death; pulmonary apoplexy 147
Maillot (drawing of) Fig. VI. 149
2. Elimination of the poison 152
3. Maintenance and increase of the strength of the patient . . . 152

CHAPTER IV.

Chronic Alcoholism.

ARTICLE I.—Intellectual symptoms 154
ARTICLE II.—Physical symptoms 156
ARTICLE III.—Cases of chronic alcoholism tending to dementia . . 160
Case XXII.—Alcoholic delirium of maniacal and melancholic form; phosphorus poisoning; tendency to dementia 161
Case XXIII.—Attacks of alcoholic delirium with epilepsy and vertigo for 20 years; weakening of intellect; tendency to dementia; hemianæsthesia 166
Case XXIV.—Chronic alcoholism with marked tendency to dementia . 173
ARTICLE IV.—Cases of chronic alcoholism terminating in general paralysis 176
Case XXV.—Alcoholic excesses; general paralysis with ataxic disorder of lower limbs; autopsy 177
Case XXVI.—Symptoms of general paralysis following excesses in alcohol; autopsy 179
Case XXVII.—Repeated attacks of alcoholic delirium; general paralysis; death; autopsy 180
Case XXVIII.—Repeated attacks of alcoholic delirium; general paralysis; autopsy 183
Sphygmographic tracings I. and II. (*general paralysis*) . . . 193
ARTICLE V.—Hemianæsthetic form of chronic alcoholism; cases of chronic alcoholism with hemiplegia and hemianæsthesia of common sensation and of the senses 198

PAGE

Case XXIX. , ... 215
Sphygmograph tracing III. (*chronic alcoholism*) 223
Case XXX. 224
Case XXXI. 229
Case XXXII. 232
ARTICLE VI.—Treatment of chronic alcoholism 233

CHAPTER V.

COMBINATIONS OF ALCOHOLISM WITH VARIOUS FORMS OF MENTAL
DERANGEMENT, AND ITS ASSOCIATION WITH INTERCURRENT
DISEASES 237

ARTICLE I.—Dipsomania 238
Case XXXIII.—Dipsomania; attempted suicide; attack of alcoholic de-
lirium 239
Case XXXIV.—Dipsomania 240
ARTICLE II.—General paralysis with alcoholic delirium . . . 241
Case XXXV. 242
Case XXXVI. 243
Case XXXVII. 244
ARTICLE III.—Insanity with delusions of persecution . . . 245
Case XXXVIII. 247
ARTICLE IV.—Imbecility; epilepsy; senile dementia . . . 248
ARTICLE V.—Alcoholic delirium developed under the influence of inter-
current diseases 249
Case XXXIX.—Attack of pneumonia in a drunkard; alcoholic delirium;
cure 252

INTRODUCTION.

ALTHOUGH for some years past, questions relating to the subject of alcoholism have enjoyed the privilege of attracting general attention, recent events have sufficiently proved that the united efforts of all, physicians, philosophers, and legislators, have not been any too great to oppose the urgent danger which threatens health, morality, and society.

In the learned discussion which was raised in the Academy by M. Verneuil, on the serious character of traumatic lesions and surgical operations on alcoholized persons, MM. Béhier and Chauffard endeavoured to shew the physical and intellectual demoralization which those who are addicted to alcoholic liquors prepare for themselves, a demoralization such that all pathological processes, whether spontaneous or traumatic in origin, present themselves under a peculiar aspect, and with characters of special gravity.

Both to surgeons and physicians, the drunkard is an exceptional subject; he feels, suffers, and reacts in a manner different from other patients.

When we regard alcoholic intoxication as a whole, we find that between the two extreme poles, that is to say, between the transient and fleeting symptoms of mere drunkenness, and the profound, permanent and irremediable disturbances of chronic alcoholism, there is a long period in which the patients even whilst presenting the general characters of alcoholic delirium, offer differences of a nature sufficiently well-marked to demand a special grouping. In one class, in fact, there are delirium, trembling, and di-

B

gestive disturbances which improve at the end
of a few days, and disappear entirely if the individual remains sober. Others shew themselves with
symptoms which quite at the onset are analogous to
the preceding, but which run a different course: in
these, in fact, the delirium persists for a longer time,
or the hallucinatory disturbances leave after their
disappearance some delirious ideas which one finds
again at the end of two or three months, and sometimes even later; these patients have usually been in
the habit of drinking to excess for a long time before, and often have already suffered from attacks
of alcoholic delirium; in them the somatic disturbances are more marked, and the intellect already
begins to shew a certain degree of weakening, yet
they are still curable, if, changing their business
or surroundings, they make a truce with their former habits.

Lastly, there is a third class of alcoholics, less
numerous than the other two, who after the attack
retain certain delirious ideas which we sometimes see
persisting for a long time after the cessation of
the physical phenomena. We have here to do with
patients of a more excitable nervous system,
whose hereditary antecedents ordinarily suffice to
explain the more potent and more lasting action
of the toxic agent. These patients, moreover, if
they get better and go out of the asylum, are not
long before they give way to fresh excesses, and one
may say that relapses are the rule in such cases.

This view of alcoholism does not, however, include all the facts. During the long succession of
pathological scenes of which it is made up alcoholic
poisoning sometimes, and, in fact, most frequently
proceeds in a slow and progressive manner, without

sudden disturbances, or interrupted only from time to time by accidents of slight intensity; at other times on the contrary, one sees phenomena of an extremely acute character break out. These very acute symptoms, which are often dependent upon recent and repeated excesses, offer, whatever be the period of the intoxication, so uniform an aspect that there is a general agreement to give them the same denomination of *delirium tremens;* of which, however, two forms are admitted, the one febrile and severe, the other apyretic and usually mild. The febrile form, a sort of *acute alcoholic delirium,* as Marcé* calls it, has its own special characteristics, and must be spoken of separately; as to the apyretic form, it presents general characters which naturally bring it into the class of simple alcoholic delirium.

Chronic intoxication, in its ultimate evolution, follows two different paths, the one ending in dementia, the other in general paralysis. These two modes of termination correspond, the first to fatty degeneration, the second to diffuse sclerosis of the nervous centres. Some cases followed by autopsy will furnish proof of this.

We shall then, in order to answer the question set by the Academy, *The various forms of alcoholic delirium and their treatment,* study in succession :

1. Drunkenness.

2. Simple alcoholic delirium, the patients being arranged in three classes according to the degree of its intensity.

3. *Delirium tremens.*

4. Chronic alcoholism, with its two terminations, the one in dementia, the other in general paralysis.

In connection with alcoholic delirium we shall have

* Marcé. *Traité des Maladies Mentales,* 1862, p. 613.

to speak of the maniacal, melancholic, and "stupid" forms, which are admitted by most authors, and which are generally in harmony with the particular characters of the hallucinations. And further, we shall have to make allowance for the symptoms produced by the abuse of absinthe.

After having determined the varied aspects under which alcoholism presents itself in its diverse forms, and the treatment which it demands in each of these forms, we shall in a final chapter, inquire into the combinations which it may offer with other diseases of the nervous centres; in these combinations, alcohol acts at times as a simple excitant and intensifies the symptoms of the primary affection without modifying them; at other times, on the contrary, affecting the diseased organism more profoundly, the toxic agent superadds its special delirium, which comes to be grafted upon the pre-existing cerebral disturbance; hence errors may arise, and in certain cases, for example, in general paralysis during its first stage, it becomes very difficult to determine the nature of the affection, this being masked by the more noisy and active manifestations of the newly added alcoholism.

Lastly, we shall have to study the influence of certain intercurrent affections, (erysipelas, pneumonia, traumatic injuries, &c.), which, by upsetting more or less suddenly the physiological equilibrium which, notwithstanding a certain degree of poisoning, the patient maintained, rapidly break down the resistance offered by the organism to the intoxication; in these new conditions, the poisoning which up to that time had remained latent reveals its existence by phenomena which are sometimes of great intensity.

ALCOHOL.

CHAPTER I.

The Immediate Action of Alcoholic Drinks— Drunkenness; The Prolonged Action—Alcoholic Delirium.

ARTICLE I.

Drunkenness in Man; Period of Excitement; Confusion of Intellect; Incoherence; Paralysis; Anæsthesia; Coma.

WHEN a man who is ordinarily of sober habits drinks to excess, he gets drunk. The first symptoms which accompany drunkenness are slight excitement, a feeling of satisfaction and of *bien-être*, in which pre-occupations are effaced, the speech and gestures grow more animated, ideas spring up more rapidly and the feelings are expressed with greater vivacity.

By degrees this excitement increases, the ideas crowd together more confusedly and are less clear, the mind begins to escape the laws of reason, and the drunken man may then appear under the most different aspects; he may be gay, sad, or full of tender emotion, without our being able hitherto to point out in these various manifestations, how much is due on the one hand, to the habitual character of the individual, his race, or the physiological conditions in which he is at the time of the drunkenness, and on the other hand, to the nature, quality, and mode of ingestion of the spirituous drinks.

Be this as it may, a thorough perturbation of the

intellect succeeds this stage of exaltation, the
ideas become confused and give rise to incoherence;
hearing and sight are not merely obscured but offer
numerous illusions; taste and smell are perverted,
and the blunting of the general sensibility sometimes
reaches such a degree of anæsthesia that the most
serious mutilations cannot overcome it.   Let us
add that the speech is thick, the countenance ex-
pressionless, the gait staggering; then the sphinc-
ters are relaxed, the functions languish, the temperature falls and the person sinks into a comatose
sleep, deprived of all feeling, and reduced as it
were to a vegetative existence.

Convulsive Drunkenness. With this ordinary drunkenness there has been ranked, wrongly, we think,
a peculiar condition designated by Percy "convulsive
drunkenness," in which there is an attack of maniacal   fury,   associated   with   clonic   convulsions.
"Ten men," says Percy, "can hardly master this
sort of madman.   His look is fierce, his eyes spark-
ling, his hair stands on end, his gestures are
threatening, he grinds his teeth, spits in the face of
assistants, and, what makes the picture still more
horrible, he tries to bite those who approach him,
imprints his nails everywhere, tears himself if his
hands are free, scratches the ground if he can break
loose, and howls fearfully."

It is plain that this is a violent maniacal outburst,
having no analogy with ordinary drunkenness, and
which we may possibly attribute to certain impure
and adulterated wines.

If illusions are frequent in drunkenness, hallucinations on the contrary are rare; certain authors,
and in particular Marcé, make no mention of them;
others, however, attributing to drunkenness symp-

toms which belong to another phase of alcoholism, mention not only hallucinations, but even ideas of suicide under their influence, maniacal impulses, &c. We have here no longer mere drunkenness, but rather acute symptoms occurring either in individuals who have for a long time been given to excess in drink, and who are consequently affected with alcoholic delirium, or else in persons who are specially predisposed on whom alcohol acts only in the character of an excitant. We shall, moreover, give an account of the development of these phenomena as obtained by physiological experiment, and by following the progressive march of the symptoms from drunkenness to the appearance of alcoholic delirium.

ARTICLE II.

DRUNKENNESS IN THE DOG.

Drunkenness and alcoholic delirium in the dog.

I. *Drunkenness; Excitement; Hebetude; Paralysis; Anæsthesia; Coma.*

What takes place when a dog is subjected to the immediate action of alcohol?

The animal, at first slightly excited, leaps, yelps, caresses, and runs hither and thither in every direction; stupor then rapidly comes on, which, at first slight, very speedily becomes more complete, and to it there succeeds a state of semi-stupor and comatose sleep, not to speak of the somatic disturbances, such as fall of temperature, anæsthesia, paralysis of the hind quarters, which gradually increases, and involves the forelimbs, caus. ing at last complete relaxation of the whole body.

Ten or twelve days later we can repeat this ex-. periment on the same animal, without exciting other symptoms, without convulsions of any sort, and especially without the appearance of illusions or hallucinations.

Now, in man, it is alcohol in the form of wine or brandy that furnishes the principal agency in the development of drunkenness; professional drinkers alone give themselves up more or less exclusively to excess in special drinks such as absinthe, vermouth, bitters, &c. But persons of this class have already overstepped the bounds of simple drunkenness. This remark is of importance by reason of the immediate symptoms which absinthe can determine, viz., hallucinatory disturbances and epileptic attacks.

II. *Alcoholic delirium; Changed behaviour; Irritability; Hallucinations; Delirium; Tremors; Vomiting; Phlegm; Gastro-enteritis.—Modes of termination; Death from cold, broncho-pneumonia, asphyxia, accident, or cachexia.*

Let us now consider the effects of the prolonged action of alcohol on the dog. From the fifteenth day of intoxication there comes on a remarkable nervous sensitiveness and susceptibility to impressions. The animal is restless, sad, he listens, is upon the watch, the least noise makes him start. As soon as the door opens, seized with great fear, crying and leaving in his path a trail of urine, he runs and cowers in the most obscure corner of the room; he no longer answers to caresses, but gets

away, hides himself, and offers to bite any one
who tries to lay hold of him, uttering sharp cries
at the least threat of a blow.

This timidity of disposition increases every day,
and after the end of the first month, illusions and
hallucinations being added, it is transformed into a
real delirium. In the middle of the night, when all
is quiet, he occasionally begins to bark furiously,
raising and multiplying his cries as at the approach
of an aggressor; at other times he groans plaintively,
it is impossible to quiet him with the voice or with
a call, one must go in with a light. At last, even
in the day time, he growls without any cause; then
thinking himself pursued he cries and runs scared
in every direction, looking behind him, and biting
in the air. These attacks of delirium are transient,
but in the intervals the animal does not regain his
habitual cheerfulness.

With the delirium there arise motor disturbances
of no less importance; the first month of intoxication has scarcely passed before one observes that in
addition to the paraplegia which follows each fresh
dose of alcohol, there is trembling in the limbs,
most marked in the hind paws, and soon advancing
to the muscles of the neck and head. This tremor,
which we have seen persist for ten or twelve hours
in the day after each fresh dose of alcohol, presents a
rhythmic type, with short and rather rapid oscillations of varying intensity, occasionally stopping
for a few seconds, and is, moreover, complicated
from time to time with quivering in small isolated
groups of muscles. Before entirely stopping the
trembling is interrupted for longer and longer
periods, until its complete cessation, after which
one can still feel muscular undulations and qui-

verings, by applying the hand to the back or the thighs.

Sensibility is not notably modified in the interval; anæsthesia occurs each time at the moment of drunkenness, but disappears with the cessation of the phenomena due to the immediate action of the poison; it is the same with the temperature, which falls 1—3° (C) according to the dose of the poison. The digestive disturbances vary according as the alcohol is given with food or is directly introduced into the stomach by the œsophageal tube, or by a gastric fistula.

In the former case vomiting is rare, the vomited matter is composed of alimentary matters mixed with mucus; the stools are generally hard and but rarely blood-stained; in the latter case, on the contrary, and especially in dogs with fistulæ, we see, chiefly in the morning on cleaning the canula, whitish, or sometimes yellowish mucus, thick, thready, viscous and streaked with blood escaping, or the animal may eject this by the mouth after some efforts at vomiting. The stools are often liquid, and occasionally black or mixed with red blood; the animal appears to suffer from colic, uttering groans and rubbing its belly on the ground. Animals thus intoxicated finish after the fashion of alcoholics; one exposed to a degree of cold of 50° Fah., already deprived of heat by the drunkenness, dies of cold; another contracts broncho-pneumonia as the result of a chill; a third is asphyxiated by the arrest of vomited food in the wind-pipe during drunkenness; another under the influence of hallucinations, escaping by a half-opened door, leaps barking from the second floor on to the flag-stones of the ground floor; and lastly, another dies out in marasmus consequent upon the alcoholic diet.

The lesions by which drunkenness is accompanied in man and animals are especially manifested in the nervous centres, the liver, kidneys, and lungs, and consist principally in circulatory disturbances, which, from slight hyperæmia rise into intense congestion and end in some cases in hæmorrhage.

M. Tardieu* has often found meningeal hæmorrhages in persons dying in a state of drunkenness; they are less frequent in animals, and this is perhaps one of the reasons why pachymeningitis consequent upon the prolonged action of alcohol appears to be rarer in them than in man.

The researches of MM. Lallemand, Perrin and Duroy† have shown, moreover, the presence of alcohol throughout the system, and its accumulation more especially in the nervous centres and the liver; other observers have confirmed most of their results, and we shall have occasion later to remind our readers of some facts in support of their opinion. As a sequel of the prolonged action of alcohol, animals present anatomical alterations, which have been previously pointed out by various authors, and which have for their basis, generally speaking, either steatosis or sclerosis.

The liver becomes the seat of a fatty degeneration, which even at the end of two months of alcohol poisoning, is already very marked; it is yellowish, strewn with points of a deeper colour. A thin section, placed under the microscope, appears slightly opaque in the situation of the more yellow points, and transparent in the other parts; the cells in the

---

* Tardieu.   *Annales d'Hygiène et de Médecine Légale.* Vol. lxix, 1848, p. 290.

† Ludger, Lallemand, Perrin et Duroy. *Du rôle de l'alcool et des anæsthésiques dans l'organisme. Recherches expérimentales.* Paris, 1860.

deeper coloured areas have lost their form, they are swollen, rounded, and infiltrated with granules and drops of fat.*

In a note recently presented to the Institute by M. Claude Bernard, M. Pupier gives the results obtained in a fowl, to which absinthe was given as a drink during a period of ten months.† "The liver," says he, "is hard, resistant, and appears lessened in volume; there are irregularities on its two surfaces, and numerous whitish depressions; the intermediate parts being of a red brown colour. Under the microscope, there is considerable dilatation of the vessels at the periphery of the lobules and they are filled with granules; extreme compression and degeneration of the hepatic cells." In a fowl subjected to the action of red wine for ten months, "the liver which is of a clear yellow colour, is soft and pasty, and oils the blade of the scalpel. Under the microscope the liver cells appear considerably enlarged and rounder than in the normal condition; they are filled with granules resembling those observed in parenchymatous inflammation at its commencement, here and there are large fat drops."

On a fowl submitted to the action of white wine, "the liver, which is of good colour, is shrivelled on its lower surface and at its borders; in microscopic sections, that which strikes one is the dilata-

---

* Magnan. *De l'action prolongée de l'Alcool chez un chien.* (Communication to the Societé de Biologie, Nov. 14th, 1868, in Compt. Rend. et Mém. 1869, p. 159).—Paul Ruge. *Influence de l'Alcool sur l'Organisme. (Virch. Arch.* xlix. p. 237, Jan. 1870.)

† Pupier. *Demonstration expérimentale de l'action des boissons dites spiritueuses sur le foie. (Compt. rend. de l'Acad. des Sciences,* May 27, 1872.

tion of the vessels, which appear of three or four times the ordinary dimensions when compared with the cells which have undergone atrophic degeneration."

On a rabbit subjected to alcohol; "Nothing as regards the capillary network, the cells, however, appear altered and contain two or three nuclei ; around the bile ducts, there is an increase of connective tissue nuclei."

" To sum up," says M. Pupier, "it appeared to us that absinthe affected primarily the stroma, without, however, producing either new connective tissue or sclerosis of the walls of the vessels; this marked new growth has not been confirmed. As for red and white wine, and alcohol, their injurious effect is produced rather in the plasma and the hepatic parenchyma."

According to M. Pupier then, alcohol would seem generally to produce hepatic steatosis, which agrees with the results obtained by other experimenters, but this by no means indicates that alcohol must inevitably produce steatosis of the liver to the exclusion of sclerosis; clinical experience would positively contradict this opinion.

On the other hand, we must notice the relatively short duration of the alcoholic intoxication in animals, compared with the prolonged poisoning of man, and one cannot positively affirm that the liver, which at first is fattily degenerated, may not, under the more continued influence of alcohol, become the seat of a more intense irritation, and as a consequence, of sclerosis. Moreover, under the same influence one may see in the same animal, fatty degeneration of the liver, irritative lesions, such as pachymeningitis, sclerosis of the posterior column of the spinal cord, thickening and opacity of the arachnoid and pia mater, milky patches in the pericardium, &c., at the same time. The alterations found in the fowl subjected to the action of absinthe do not appear to be either sclerosis or steatosis, and the fact of these changes resting only on the authority of a single case should, as M. Pupier himself says, cause us to suspend our judgment. For our own part, on a dog which had a gastric fistula, and to which we gave from 60 to 120 grammes of absinthe liqueur every day for a month, we did not at the autopsy find any lesion in the liver which could remind us of that of which M. Pupier speaks.

The kidneys, like the liver, undergo a commencement of fatty degeneration, the surface remains smooth and even, the cortical substance and the prolongation between the pyramids of Malpighi show a well-marked yellowish tint, with small striations of a deeper yellow colour; under the microscope one sees the *tubuli* slightly swollen, a little cloudy and filled with granular and fatty epithelium.

M. Ruge mentions adhesion of the capsule to the renal substance in four cases. In three cases the same observer has noted fatty degeneration of the heart. For our part we have several times seen evident traces of pericarditis. The coats of the stomach are not sensibly thickened in dogs which take alcohol mixed with their food; the mucous membrane is injected along the greater curvature but very rarely ulcerated. Those on the contrary in which, apart from meals, alcohol is conveyed directly into the stomach either by the æsophageal tube or by a fistula present traces of a violent gastritis. In the case communicated to the Societé de Biologie (on Nov. 14th, 1868), the stomach was found to be shrivelled and thickened, and in addition, "the surface of the reddish brown mucous membrane was lined with a layer of thick, sticky, glairy mucus, streaked with blood; on cleansing these parts by a stream of water, one saw beneath, small ulcerations with irregular borders, and in some places cicatrized portions in the form of irregular greyish plates. In the substance of the mucous membrane were found infiltrations of blood, some spread out in a layer, others united into small spots." In this same dog "the cord is not injected, and its consistence appears normal throughout its whole extent. On making sections in different regions, one sees a greyish tint in the posterior columns, which is hardly visible in the upper part of the cord, but becomes more marked towards the lower third, where it has the form of a triangle with the base directed backwards on each side of the posterior median fissure. In this region, one sees also a slight greyish tint in the anterior column on each side of the commissure." We shall have occasion further on to cite analogous cases in man, when speaking of the termination of chronic alcoholism by general paralysis.

Pachymeningitis as the result of the prolonged action of alcohol has been noticed by Kremiansky in dogs which had been made to swallow considerable doses of alcohol during four weeks.* M. Neumann (of Königsberg) has had the opportunity of observing the same fact, after an alcoholic intoxication likewise of four weeks, but in other cases he did not find any alteration whatever in the meninges of ten dogs submitted to daily drunkenness for several months;

---

* Kremiansky. *On Internal Hæmorrhagic Pachymeningitis in Man and the Dog.* (*Virch. Arch.* xlii. p. 129, 1868.)

on the other hand we ourselves have found slight
infiltration and a little thickening of the arachnoid
and pia mater, but without false membranes on the
dura mater.   M. Ruge, for his part, has only observed in the dura mater dilatation of the vessels, and in
the pia mater simply injection or œdema.

Whence arises this diversity in the results obtained
by analogous experiments? The cause seems to
us to reside in the production of meningeal hæmorrhage in some animals during the drunkenness.
Without wishing to raise here the interesting question of meningeal hæmorrhages and consecutive false
membranes, so often discussed since the time that
M. Baillarger determined their precise seat in the
arachnoid cavity,* we will remind our readers that
M. Laborde by puncturing the superior longitudinal
sinus in young dogs obtained extravasations of
blood in the arachnoid cavity, which were surrounded by a membrane of new formation as early
as the tenth day.   More recently M. Sperling,†
operating on the rabbit, injected fresh blood into
the arachnoid cavity.   Eight days after the operation, one observes the commencing formation of
a new membrane which becomes completely organized at the end of two or three weeks; in all
cases, one can at the end of three weeks find a
new formation of vessels.   These new membranes
are entirely analogous to those which one finds in
hæmorrhagic pachymeningitis.   The organization
of the blood over the concavity of the brain and on

---

* Baillarger, *Du Siége de quelques hémorrhagies méningées*, 1837.
Laborde, *Sur la pathogénie des hémorrhagies méningées intra-
arachnoidiennes.   (Société anatomique*, 1864.)

† Sperling, *Expériences sur la pachyméningite hémorrhagique*
(Centralblatt, 1871, analyzed in the *Gaz. hebd.* Jan. 12th, 1872.)

the internal surface of the dura mater takes place without adhesions to the arachnoid.

At the meeting of the Société de Biologie, on April 6th, 1872, M. Vulpian in relating some experiments made with M. Philippeau mentioned the encysting, by a false membrane, of blood taken from the femoral of a dog and injected at once into the pleural cavity of another dog. Injections of blood into the arachnoid, have also led similarly to a rapid formation of false membranes, and the clot thus developed has remained attached to the dura mater.

All these facts tell in favour of the interpretation which we would give of the existence of new membranes in some alcoholized dogs; but we must not lose sight of the fact that pachymeningitis may come on without pre-existing hæmorrhage in the course of certain affections of the nervous centres, such as general paralysis, senile dementia, and chronic alcoholism, diseases in which the nervous system in its whole extent may at times become the seat of a very intense irritative process.

To sum up, physiological experiment, after allowing us to watch the successive development of intellectual and physical disturbances, shows us anatomical changes in the nervous centres, in the alimentary canal, the liver and the kidneys, and in the respiratory and circulatory systems, which it is the more useful to determine precisely, since they represent the earlier stages of the more profound disorders to which the human organism succumbs in the later periods of intoxication.

We relate here one of our experiments to give a more complete idea of the prolonged action of alcohol in animals.

EXPERIMENT I.—Dec. 1st, 1869.—On a terrier dog, two months old, vigorous, weighing 6 kilogrammes (13 lbs.) whose temperature in the rectum on the evening of the preceding day was 39·2° C. (102· 6° F.)

The animal takes with its food 25 grammes (nearly 1 oz.) of alcohol, the common spirits of commerce : at the end of eight minutes, and whilst with its head lowered it is finishing the scraps, it is seen to rock on its legs from before backwards, and losing its equilibrium ends by falling on to the dish ; then it gets up, and walks staggering, straddling its legs, which bend first to one side, then to the other, but especially backwards. Sensibility is diminished but not abolished ; striking or squeezing the tail provokes faint cries. When placed on its litter the animal speedily goes to sleep ; at the end of half an hour it wakes when called, rises and walks in a staggering manner. Temperature in rectum, 38·2° C. (100·8° F.)

From Dec. 2nd to 8th. Is given each day 25 grammes (1 oz.,) of alcohol and the same symptoms are reproduced.

From 9th to 11th. Rest; diet without alcohol.

From 12th to 18th. Daily dose of 30 grammes. Production each day of drunkenness with staggering, slight anæsthesia and lowering of temperature.

From the 19th to 22nd. Daily dose 30 grammes. The animal becomes unhappy, restless ; the slightest noise makes him start ; as soon as one enters he moves and hides himself, does not come when called, trembles if one approaches to seize him, cries or tries to bite when touched. After the meal, the drunkenness and stupor last from 3 to 5 hours and are followed by a nervous excitability which becomes more marked every day.

From the 22nd to 28th. Daily dose, 30 grammes. The restless condition of the animal increases, he appears constantly in a state of terror ; in the night when all is quiet, he utters groans, and barks with unwonted energy ; at times is heard running about yelping in different parts of the room; on knocking at the door, or calling him, he is quieted for an instant, but soon begins again ; if one goes in with a light, he is reassured, but soon returns to his frightened state. In the day, he yelps at times, pricks his ears, listens and growls, but is less excited than at night.

From the 26th to 27th. Has little appetite, and takes longer to eat his meal.

29th. Diet without alcohol ; attacks of fright, especially at night.

From Dec. 30th to Jan. 15th, 1870. Daily dose, 40 grammes (nearly one and a half ounces.) The illusions and hallucinations occur more frequently during the daytime; he is seen to start up all of a sudden, bark furiously and run away with his head down as if to avoid blows, then turning back and biting in the air. During the attack of fright, the urine several times escapes in jets, and sometimes even gas and fæcal matters are suddenly expelled. The temperature taken during these delirious attacks, which habitually occur apart from the drunkenness, does not vary from the normal. We obtained, always using the same thermometer from the beginning of the experiment, the temperatures 38·8° C. (101·8° F.) 39° C. (102·2° F.) and 39·2° C. (102·6° F.) During the drunkenness following the meal, we continue to find a slight lowering of the temperature varying between 38·4° C. (101·1° F.) and 37·8° (100° F.) On Jan. 3rd, the rectal temperature fell to 37° C. (98·6° F.); we had given with the alcohol some remains of fish, the animal had eaten more greedily and the muscular relaxation had been more complete and more rapid.

At this period of the experiment phenomena of another kind are noted with regard to the motor functions; besides the staggering and the paralysis which occur each day during the drunkenness, trembling begins to appear in the paws; when the animal is stretched out one perceives rhythmic shocks with short and rapid oscillations, more marked in the hind limbs; this trembling is irregularly intermittent, it stops for some minutes and then recurs again with the same characters; it usually lasts for three or four hours and ceases after the drunkenness.

Jan. 16th. Diet without alcohol.

From Jan. 17th to 31st. Daily dose, 40 grammes. The delirium with illusions and hallucinations continues to be produced night and day; on several occasions, the animal presents a slight excitement at the onset of the drunkenness, is more cheerful, comes quickly when called, and is affectionate; then reels and becomes stupified; after the drunkenness, returns to the same timid and frightened state as at first. The trembling increases in extent, duration, and intensity; the muscles of the back and shoulders give slight shocks which can be felt by the hand. On two occasions vomiting occurs.

February. Daily dose, 50 grammes (nearly 2 ounces.) During the month, four days of rest with non-alcoholic diet. The animal has less appetite, and eats nothing with relish except fish.

The stupor is more prolonged after the drunkenness; the delirium appears less active; the nights are less disturbed; the animal is only heard to groan and cry three or four times; during the day the attacks of fright and anger are also less frequent and violent.

The trembling, which is more marked, persists after the transitory phenomena of the daily drunkenness; from the paws it gains the trunk and head, the latter oscillating in a very singular manner as in patients affected with paralysis agitans. The trembling, however, although much more prolonged, does not last the whole day. The animal has become thin, the hair dull, the eyes often bleared. The beats of the heart are weak. Vomiting comes on at five different times during the course of the month; constipation is habitual, and at times streaks of blood colour the excrement.

March. Daily dose, 40 grammes. Four days rest in the course of the month. Appetite lessened, great thirst; vomiting only twice. The same degree of trembling; stupidity; sadness; less delirium. The urine does not give signs of albumen or sugar on examination.

April. Daily dose, 30 grammes. Five times, a rest of twenty-four hours with non-alcoholized diet.

The animal is more and more miserable, stupified, sometimes passes several hours in eating his meal; the trembling does not increase, the delirium remains the same.

May. Daily dose, 30 grammes. The animal eats little, will hardly touch meat now; takes, however, all his meal when fish is given; on the 14th ate with greediness a large quantity, and swallowed about 55 grammes of alcohol mixed with his food, and then fell into a drunken comatose condition; during the night goes out of his kennel, no doubt still staggering, and lies stretched out on the ground by the side of a vase from which part of the water has been spilt. In the morning is found wet, lying at full length, short of breath, with the ears cold; pulse 130, respirations 35, temperature in rectum 40·2° C. (104·4° F.) : on auscultation, harsh breathing and subcrepitant râles at the base of the right lung, disseminated râles in the remainder of the chest.

15th. Reddish viscous discharge from the nostrils, ears warm. T. R. 40·6° C. (105° F.) The animal will take nothing, but lies· stretched out upon its side.

16th. Symptoms aggravated; very great difficulty of breathing, extreme weakness.

17th. Died in the morning.

Autopsy. The cerebral dura mater is slightly injected, but does not show false membranes; the arachnoid and pia mater, which are œdematous at the base of the brain, have a rosy tint over the interpeduncular space; they separate easily everywhere. Sections of the hemispheres show a fine stippling, but there is no distinct collection of blood. The surfaces of the ventricles are injected, abundant vascular ramifications spread out upon the upper surfaces of the optic thalamus and corpus striatum, where the ependyma is slightly thickened; on section, nothing is found in the deeper parts.

The spinal dura mater appears normal, but the arachnoid and pia mater are injected, especially at the lower part of the dorsal region. Sections of the cord shew a notable injection of the grey matter, but no gelatinous tint is visible in any portion of the columns. Microscopical examination does not discover any alteration.

The lungs present in places large marblings, passing from pale rose to a blackish red colour; they look like the lungs of a child which has died from broncho-pneumonia. The deep red portions are dense, resistant to the touch, non-crepitant, and sink in water. On the right lung at the base, section shews some greyish points. On following the divisions of the bronchi with scissors, they are found loaded with thick viscid mucus mixed with blood. The marginal portions of the lobes have remained rosy almost everywhere, slightly prominent at the situation of some emphysematous lobules, the blackish parts on the contrary are depressed at the surface. The right cavities of the heart are distended with liquid blood mixed with black clots; the left present no particular change. The visceral pericardium has an opaline tint, and is milky over the coronary arteries, especially at the base.

The liver is yellowish, strewn with deeper coloured points, which, when examined under the microscope, are the seat of an already well marked fatty degeneration.

The kidneys are yellowish in the region of the cortical substance and its prolongations between the pyramids. The spleen appears normal.

The stomach contains several small clots; the surface of the mucous membrane is covered with a viscid, very adherent mucus; the mucous membrane is thickened but does not shew any ulcerations. (*Archiv. de Physiol.*, May, 1873).

It remains for us to call in the aid of experimental physiology to clear up an interesting feature

of intoxication by the abuse of spirituous liquors ;
I speak of the action peculiar to absinthe.

ARTICLE III.

*Action of absinthe on animals ; Muscular shocks.   Vertigo ;
Epileptic attack; Hallucinations; Delirium.   Action of
absinthe in man.*

Even a long time ago, authors had recognized
the properties peculiar to absinthe.   MM. Trousseau
and Pidoux, in their treatise on Therapeutics say,
in speaking of its action, "It is certain, that the
liqueur known under the name of water or cream
of absinthe intoxicates very readily, and produces
vertigo and a state of nausea which, in this case,
is not due to the alcohol but to the absinthe.*   M.
Bouchardat and M. Gubler also recognize in this
substance qualities distinct from those of alcohol.   M.
Meynier,† comparing the essence of absinthe with
essence of camomile ascribes to it an antispasmodic
and anæsthetic action, an action which, as we shall
see, is very different from that which it really
possesses.   But it is especially since 1864, after the
first researches made with our lamented master
Marcé at the Bicêtre, that the toxic properties of
absinthe have been determined with more precision.‡   Other works have successively come to
throw some light upon this question, to confirm some

---

* Trousseau and Pidoux, *Traité de thérapeutique et de matière
médicale.*   Vol. ii., p. 494.

† Meynier.   *Recherches sur l'action toxique de quelques essences.*
(Thèse de Paris, 1859).

‡ Marcé.   *Note sur l'action toxique de l'essence de absinthe.*
(*Compt. rend. des séances de l'Acad. des Sciences,* 1864, vol. lviii., p.
628.)—Magnan.   *Accidents déterminés pas l'abus de la liqueur d'absinthe.*   (*Union Medicale,* Aug. 4th and 9th, 1864.)

of the results already obtained, and to fill up the gaps left by the earlier experiments.*

Whatever be the path traversed by the poison to reach the nervous centres, what one observes is as follows; when weak doses of essence of absinthe have been absorbed, one sees after a more or less marked muscular tremor, little abrupt jerking shocks like electric discharges, which are repeated once or several times in the muscles of the neck, and give rise to rapid and very limited movements of the head, which is drawn upwards and backwards; the contractions gain successively the muscles of the shoulders and the back, and then excite abrupt shocks which raise the fore part of the body by jerks from the ground; the animal crouches, and rolls itself up, and seems to resist with all its might these powerful discharges. This fact of the action of absinthe being directed more especially to the head and neck, that is to say, its influence on the medullary and cervical regions of the cord, is the more remarkable as alcohol acts in a directly opposite manner; the latter in fact, brings on paraplegia first before paralysing the anterior parts of the body, absinthe, on the contrary, excites muscular shocks in the fore quarters before producing general convulsions. Nor is this all, in some cases one sees a very interesting phenomenon come on in the dog; the animal stops all of a sudden and

* R. Amory. *Experiments and reflections on absinthe and absinthism. (Boston Med. and Surg. Journ.,* March 5th and 12th, 1868.)— Challand. *Etude expérimentale et clinique 'sur l'absinthisme et l'alcoolisme.* (Thèse de Paris, 1871.)—Magnan. *Etude expér. et clin. sur l'alcoolisme &c., épilepsie absinthique,* 1871.—*Rech. de physiol. path. avec l'alcool et l'essence d'absinthe.—Epilepsie.* (*Arch. de phys.* March and May 1873).

stands motionless, as if stupified, with head down, expression gloomy, and tail lowered; he keeps this attitude from 30 seconds to 2 minutes, and then spontaneously recovers his usual condition. This is a state of giddiness which is not without analogy with *petit mal* or the "absence" of epilepsy.

The action of essence of absinthe in a large dose is different, or rather it is a higher degree of intensity of the same phenomena. After premonitory symptoms resembling those of which we have just now spoken, or else quite abruptly and more or less quickly, according to the mode of introduction of the poison, the attacks come on, the animal falls suddenly with trismus and tonic convulsions, which constitute the first stage of the convulsive seizure. Most frequently the extensor muscles of the neck and back contract energetically so as to produce opisthotonos, but with it one almost always sees some pleurosthotonos; the muscles of one side of the body contract and produce a bow-shaped curve in which one side rests on the ground by its middle part, whilst the two extremities are raised and tend to curve so as to approach the opposite side. In some rare cases, it is the flexor muscles of the neck and trunk which go into convulsions first, and in a dog who had taken four grammes (1 drachm) of essence of absinthe, the flexion was so pronounced at the onset of the attack that whilst the head was applied to the front of the chest, the posterior part of the neck, the shoulders and the trunk touched the ground in succession, and the animal before falling, turned once over himself in the most marked emprosthotonos. In a second attack, the head which was at first flexed was rapidly carried into extension; three other very violent attacks occurred in succession with extension of the head during the tonic stage, and the animal died in the fifth attack.

To these tonic convulsions there succeed at the end of a few seconds clonic convulsions with snapping of the jaws, which clash or else are brought convulsively together without actually coming in contact; foaming at the mouth and sometimes biting of the tongue supervene; stertorous breathing, evacuations of urine and fœcal matter and even of semen follow in some cases.

The attack over, the animal usually remains slightly stupified and then comes to himself again; the attacks of epilepsy sometimes occurring with this typical course of symptoms, with intervals between them of ten or twenty minutes or even longer. In these cases, which are of more constant occurrence when the poison has been introduced into the stomach, one can give a very clear idea of the phenomena; but this is not always the case; some symptoms may be wanting, or they may succeed each other very rapidly and be confounded together. In some cases the attacks follow each other without interruption, and form composite, confused attacks, with violent and rapid convulsive phenomena, in the midst of which one distinguishes less clearly the attack of epilepsy.

Such are the convulsive phenomena due to absinthe, very different both from the paraplegia which accompanies alcoholic drunkenness and from the trembling which follows the prolonged use of alcohol; vertigo and attacks of epilepsy are, in fact, the two degrees of acute poisoning by absinthe. But, in addition, one sees sometimes disturbances of intellect which are very important in their bearing on the question before us. Whilst alcohol, as we have seen, requires to prepare the ground in order to give rise to hallucinations, absinthe excites the phenomena of hallucination at the very outset.

EXPERIMENT II. *Hallucinations and epileptic attacks in a dog under the influence of essence of absinthe.* In a dog weighing 31 lbs. into whose stomach we injected 5 grammes (about 75 minims) of essence of absinthe at 9·15 A.M., we observed a first attack of epilepsy at 9·45 A.M.; ten minutes later a second attack occurs followed by a slight degree of stupor; quickly becoming himself again, the animal continues playful, answers a call, walks and runs easily. Quite suddenly and without any provocation, he raises himself on his feet, with hair bristling, angry look, eyes injected and brilliant; he fixes his gaze on a wall which is completely bare and on which there is nothing to draw his attention, bending down with the paws forward, and the neck stretched ready to spring, he advances and recoils alternately, barks furiously and gives himself up to a furious battle; clashing his jaws, and making sudden movements as if to seize an enemy, he shakes his head from side to side clenching his teeth as if to tear his prey. By degrees, he becomes calm, but still looks several times in the same direction, growling, and then regains confidence completely.

I give also another experiment borrowed from the thesis of M. Challand* in which the hallucinations were developed with some degree of intensity.

EXPERIMENT III. *Hallucinations and epileptic attacks in a dog under the influence of essence of absinthe.* An injection of 4 grammes (nearly 1 drachm) of essence of absinthe was made into the stomach of a young bull-dog, at 9·30 a.m. The rectal temperature was taken immediately, and was 39·8° C. (103·6° F.).

At the end of four or five minutes the animal seems restless and agitated, groans and whines a little. Respiration becomes frequent and there is foam on the mouth. At the end of 20 minutes some very slight shocks were observed, or rather a mus-

---

* Challand. *Loc. cit.* p. 13. Experiment made at the Hotel Dieu at Prof. Behier's clinique.

cular trembling in the ears and at the nape of the neck. He displays great nervous excitability, starts at the slightest noise, and gives signs of dread.

In order to obtain a very rapid effect, a quarter of an hour later 3 grammes more of the essence were injected into the stomach; ten minutes were allowed to elapse, and as the animal had still only slight shocks, the enormous dose of $1\frac{1}{2}$ grammes of the essence was injected into the crural vein.

Five minutes after, there was an abrupt movement of recoil accompanied by slight convulsive movements of the head and trunk, and then he fell suddenly on to his side. All the limbs were stiffened, and stretched out convulsively, the jaws strongly clenched, the body raised up as it were in the middle: the tonic convulsions lasting from 15 to 20 seconds, and succeeded by very prolonged clonic convulsions. The animal continues to lie on the side, but is no longer raised up, the limbs are agitated by strong shocks which make them move backwards and forwards very rapidly. The breath becomes gasping and then stertorous, and the teeth chatter.

An abundant frothy foam flows from the lips. The rectal temperature is $39.8°$ C. ($103.6°$ F.).

There are also during the convulsions evacuation of urine and of fæcal matters. The clonic convulsions last from two and a half to three minutes, and then cease for a short time (about five minutes), only the very gasping and stertorous respiration, and the flow of thick frothy foam continuing. Then a fresh attack begins, which is of very violent character, and is followed by another of equal severity after a very short pause. The two attacks are in every respect similar to the first and are followed by another series of attacks which are so close together that they run into one another. In these last the tonic convulsions are much weaker and are immediately replaced by clonic convulsions. The eyes are injected, the pupils widely dilated. From time to time a small quantity of urine is discharged, T. R. $39.8°$ C.

Since the first attack about half an hour has now elapsed. By degrees the convulsions cease and the slight convulsions of the commencement reappear. The animal seems to be coming to himself again a little. The pupils continue dilated, the eyes haggard, and he looks constantly in one direction, and wants to bark, but can only make some indistinct sounds.

Then he tries to stand, but cannot succeed at first, and as he recovers consciousness other phenomena shew themselves. He has very distinct hallucinations, suddenly barks furiously, although he is not touched or excited, growls constantly and tries to bite some imaginary thing; his jaws only clashing in the air. If objects are set before him, he tries to bite them, but cannot succeed. He ends by getting up entirely; but the hallucinations continue.; when a slight noise is made in the room, he becomes furiously enraged and wants to rush in one or other direction; a little later still (11.30 A.M.) the hallucinations become more frequent, but in the intervals perception is clearer, he turns when called, he is however still weak on the legs and has a vague look.

During the day the animal completely recovered himself; only a slight swelling in the thigh remaining, where the injection into the crural vein had been made.

On July 3rd, that is to say, four days afterwards, no symptoms remained, but still later, symptoms of another kind appeared in the chest (infarcts, embolism of the pulmonary capillaries, pulmonary apoplexy, &c.)

M. Challand points out the considerable quantity of absinthe injected in this dog, 7 grammes ($\frac{1}{4}$ oz.) in two injections into the stomach, and $1\frac{1}{2}$ grammes (23 minims) into the crural vein, whilst ordinarily a dose of 15 to 20 centigrammes (3 minims) injected into the veins is sufficient to produce the symptoms of the poison. He adds, with reason, that the essence was of inferior quality, and came from a bottle which had been left uncorked. This, in fact, is what usually happens, when the precaution is not taken of closing the phial hermetically; but it must be added that the activity of the different essences employed in commerce is variable, some being obtained by the distillation of fresh plants, others of dried ones.

Further physiological researches, aided by chemical analysis of these substances, may fill up the gap, by determining what are the active ingredients of the complex product which is known by the name of absinthe. But, for the present, the results ob-

tained are sufficient for our study, since the essences employed are precisely those usually employed in commerce for the liqueurs.

The fact of the premature appearance of hallucinatory disturbances under the influence of absinthe
had not escaped a sagacious observer, M. Motet,
who as early as 1859 had pointed out as a clinical
fact, that delirium developed more rapidly in drinkers of absinthe. " In these cases," says he, " there
are found no doubt all the general characters of
alcoholic delirium ; but that which lends a special
character to intoxication by absinthe, is the total
absence at this stage (viz. the acute) of muscular
tremor ; it would appear as if one had under observation a form which had been, as it were, hatched
before its complete development, and in which the
rapidity of action of the cause · had prevented the
production of the usual phenomena.*

Of the two active agents in the liqueur absinthe,
the alcohol and the absinthe, the latter poison had
produced the delirium and the hallucinations before
the alcohol had had time to exert a sufficient action
on the nervous centres to induce trembling. This
is what we learn from the cases nos. 1 and 2 of M.
Motet,† especially the first; the individual, habitually sober, had begun drinking only a few days
before, and took from ten to twelve glasses of absinthe daily. But the case is sufficiently important
to be reproduced at full length.

CASE I.† Jean Baptiste S., aged 34. This man, who was of

---

* Motet. *Considérations générales sur l'alcoolisme et plus particu-
lièrement des effets toxiques sur l'homme par la liqueur d'absinthe*, 1859.
p. 18. (In cases i. and ii. the convulsive attacks are not in question ; they are noticed in case v.).

† Motet. *Loc. cit.*, p. 31.

middle height, and habitually good health, had worked for three months as a sculptor at the decoration of the exterior of the Louvre. He was steady, sober, and a hard worker; when all at once, fifteen days ago, yielding to the inducements of his fellow-workmen, he deserted his work for the wineshop. He soon got to drink as much as 10 or 12 glasses of absinthe a day, and then brandy, white wine, &c. Some days before the outburst of the delirium, he still drank, but without being able to get tipsy, he was "as if stupified." In the night of the 29th to 30th of May, he jumped out of bed suddenly; he saw gendarmes who were coming to arrest him. He protests his innocence, and then remains seated without saying anything until daylight.

He is a little more calm towards morning; goes out, drinks some absinthe, and comes in in the evening in a state of complete stupor. He is helped to bed, but is hardly in bed before terrifying visions begin to assail him; he cannot close his eyes for a moment, he gets up and walks about the room, sees enemies in the persons who are placed to take care of him; and as his condition gave rise to some anxiety, steps were taken to place him in a private asylum on May 31st, 1857.

As soon as his antecedents were known to us we thought that we had to do with a case of *delirium tremens*. On examining the patient we were astonished not to find in him any staggering, or trembling of the hands or lips. On our arrival he was seated, his face had an altered anxious look, and remained motionless; his attitude was rather that of a melancholic than an alcoholised patient. Nevertheless one could see that there was something peculiar in the expression, which had retained all its vivacity; in the warmth of the skin, and even in the rapidity of the movements of the head, which was turned from side to side to notice what was going on. There was not the concentration of melancholia with stupor, nor the fixed look, and the immobility which characterize it. The pulse was full, of moderate frequency: (90) the tongue thickly furred: the pupils not much dilated.

He answers the questions which are asked him, but with great indecision. He does not know "if he is really himself; his name is S——, he does not understand why he is called Jean. It is not he, he has not killed anybody, why is he insulted?" He listens, turning his head every moment to the side whence the strange sounds come to him. He walks with a fair amount of confidence, but his legs are rather weak.

Absolute insomnia during the night; he does not cry but speaks in a low voice, sees flames surrounding his bed, and grinning faces standing out in relief on the walls. He is pursued by gendarmes. No sights of dirty animals.

June 1st. A bottle of Seidlitz water, two dishes of soup only, and two litres of vegetable lemonade. Condition the same during the night and day ; motions free.

On the following morning, he is put into a bath, and remains there for five hours, cold affusions are applied to his head, and they are repeated every half hour. We see the patient in his bath, and he seems to us less excited. This improvement was, however, only apparent. Two hours after this bath the same ideas existed. He doubted his own identity, affirmed that he had not committed any crime, &c. If one asked him for any information about his business he gave it very exactly, but with very little detail, distracted as he was by accusing voices. Beyond this there was no agitation, no cries, and no disordered movements.

The prolonged baths were continued for two days longer, and then on the following day, another purgative (Seidlitz water).

The succeeding night was much more calm ; S—— slept nearly two hours without any waking in starts.

On May 7th, we find him walking in the garden, still a little restless ; he asks us what we want to do with him ? if he could not see his father ? He listens to us with a more sustained attention. He has a sort of half consciousness of his condition, feels vaguely that it is not all right, but does not know what has taken place. (Prolonged bath, cold affusions). In the evening, S—— dines with more appetite than on the preceding days, and sleeps with calm slumber for nearly five hours.

On the 10th, we find him better ; he has still some disquieting preoccupations, but he makes inquiries about his family. Has no precise recollections, is ignorant of how long he has been in the establishment. We tell him that he has been ill in consequence of alcoholic excesses, and he admits irregularity of his mode of life, and the change which had come over his habits, but he still is afraid of being arrested and thrown into prison.

From the 10th to the 15th. The appetite becomes regular, the digestive functions are normal, the delirious ideas disappear, and we consider S—— cured.

He went out on June 25th perfectly recovered.

This observation of M. Motet has the greater value because it relates to a patient of regular life,

and sober up to the moment when particular circumstances led him to drink absinthe.   Thus, freed
from all complication, the clinical fact acquires the
value of an experiment in which the subject of it is
placed in precise and clearly determined conditions.
If the hallucinations and delirium were only produced at the end of several days in this patient,
whilst they may occur immediately in the dog, it is
because the latter takes not merely a larger dose of
poison, but the pure essence of absinthe without the
addition of alcohol, which, as we know, retards but
does not prevent the action of the essence of absinthe.*

But if the hallucinatory disturbances differ in
their mode of appearance according to the agent
which gives rise to them, whether alcohol or absinthe, their general characters are the same in both
cases.   These hallucinations are painful, disagreeable and aggressive, they have on the other hand
a mobility which the expressive gestures, the atti-
tudes and the rapid movements of animals—and one
may say also of men, sufficiently exhibit.

* Magnan. *Loc. cit.*, p. 28.

CHAPTER II.

ALCOHOLIC DELIRIUM IN MAN.

JUST as the animal, so does man, after a little while, change in character, appear irritable, restless, and easily impressed; he no longer sleeps; he becomes the sport of illusions and hallucinations, and when, after repeated excesses, he passes the limit of saturation, or is subjected to any other cause of excitement, he is seized with an attack of alcoholic delirium.

ARTICLE I.

General Characters of Alcoholic Delirium.

I. *Painful character of the hallucinations.*—The intellectual phenomena consist especially in hallucinatory disturbances, which are exceptionally of a gay nature, almost always, on the contrary, as has been long ago noticed, of a painful character, giving rise to fears of every kind, and as M. Marcel says, capable of giving rise to moral impressions, of which the slightest would be astonishment, and the strongest a profound terror.[*]

The alcoholic Jean J——, (Case viii.) seemed as if he must have been an exception to this rule at the times when he spoke of cascades which he heard roaring, illuminated chalets which he saw, and songs and dances of which he was witness. But even from these pleasant sights, these songs and feasts, he could create the saddest delusions. The cascades

---

[*] Marcel. *De la folie causée par l'abus des boissons alcooliques.* (Thèse de Paris, 1847, p. 12).

D

impelled him, he said, to drown himself. " That
sound of water was to defy me, it meant that I
was a coward, and that I had not the courage to
throw myself into the water." The chalet with
its joyous occupants. "that was a lot of people
who wanted to destroy me, who were mocking at
me and wished to assassinate me."

Some patients, however, have at times agreeable
hallucinations, they hear music, and melodious
voices; they see landscapes, or charming pictures;
one such asserted that he smelt the most exquisite
perfumes. An old soldier who had gone through the
campaigns in the Crimea and Mexico, and who had
spent several years in Algeria, whilst in an attack of
alcoholic delirium, saw magnificent forests with immense trees; groups of young girls, singing and
dancing, crossed groves of flowers lighted by lamps
of a thousand colours ... but in a few moments, the
picture grew gloomy, and lions, hyænas and serpents replaced the flowers and dancers.

II.   *The changeable character of the hallucinations.*—
Another character of the delirious conceptions and
illusory perceptions of alcoholics, as has been well
shewn by M. Lasègue, is their mobility.* Men,
things, and animals, in fact everything which forms
the subject of the hallucinations, moves and changes
its place; hence the changefulness and rapidity of
the ideas and acts of the alcoholic, who himself,
moreover, either terrified, anxious, restless, suppli-
ant, or aggressive, is always engaged in the most
active manner.

III.   *The hallucinations have for their subject
either the ordinary avocations, or the dominant subjects
of interest of the moment.*—These hallucinations

* Lasègue.   *De l'alcoolisme subaigu.   Arch. gén. de méd.,* 1869.

vary in an infinite degree, but they are often a
reflection either of the daily occupations, or of the
predominant pre-occupations of the time, with, however, a marked preference for whatever is the most
painful and disagreeable. Thus, at the time of
recent events, during the German war, the patients
saw Prussians, heard themselves styled spies, they
were called Bismarck; later, during and directly
after the Commune, the alcoholics belonging to
the regular troops saw Communists and National
Guards, whilst the latter on the contrary saw
Versailles troops. Under the Empire, it was especially spies and policemen who were seen; and
at all times thieves, celebrated assassins, the Tropp-
manns of every kind have furnished the material of
the hallucinations. When the hallucinations relate
to the habits or professions of the persons, they
sometimes make the patient act his part in the
most singular manner. Thus the patient (Case
iii.) who was a costermonger, saw around him in
the ground, cauliflowers, sorrel, and radishes, which
he made great efforts not to crush under his feet;
a drover urged on his dog, and saw and called cat-
tle and sheep; the man who fed pigeons at the
market (Case ii.) thought he held a pigeon between
his fingers and tried to make it swallow corn;
a winedealer answered her customers, asked them
to wait and prepared to serve them; the carpenter
(Case xx.) saw the planks with which he was trying
to load a cart, fall on his head and back. A street-
walker had obscene ideas and saw around her wan-
ton scenes which brought to mind her unhappy
calling.

These hallucinations, according to their intensity,
and according also to the disposition of their sub-

ject, give rise to different reactions, which have the power of completely changing the aspect of the patient. Hence arise the maniacal, melancholic, and stupid* forms of alcoholic insanity, forms which might be multiplied, though without any real advantage, if one wished to express clearly all the varieties under which the disease may present itself.

The maniacal form is the most frequent, and we find it noted in most of our observations; it sometimes precedes or follows the two other forms in the same patient, without altering in any way either the general features or course of the disease.

The patient Louis D—— (Case xii.) was in a state of complete mania on the day of his arrival at the Asylum, crying out, gesticulating, and moving about everywhere; on the next day he had the appearance of a melancholic with partial stupor; he was sad, pre-occupied, and did not answer questions; he looked frightened, and at times shook his head sadly, held his hands firmly over the upper part of his thighs, got up and cast uneasy glances around him, and then sat motionless on his chair. Some days after, he was better, his countenance open, he talked willingly, and then explained the strange fears which tormented him, and which accounted for his attitude. He thought he was in prison, before a tribunal, accused of shameful acts, and believed that he was condemned to suffer a horrible mutilation; he felt the instrument cutting off his genitals, whilst mocking voices bitterly jeered him.

---

* "Stupidity" is a word used by French alienists to indicate the form of insanity usually known as "melancholia with stupor," and is defined by Griesinger as melancholia in which the patient is lost in self-contemplation. Griesinger. *Mental diseases,* Syd. Soc. Trans., p. 246.

The patient Anastasia D—— (Case iv.) was very excited on entering the Asylum; covered with sweat, moving incessantly, she moved her hands to different parts of her body as if to push away or seize animals or insects; stretched out her arms in the air, pursued imaginary beings, answered, cried out, called, wanted to open the doors and windows, hid herself, defended herself, &c., in fact was in a typical maniacal condition. During the night the scene changes; the patient will not stir, will answer no question, stays motionless, allows herself to be moved about, and makes no movement to rid herself of the saliva which fills her mouth and flows over her lips, passes her water under her, and remains in a complete state of stupor. The next day she presents rather the form of melancholia with slight excitement.

But how comes it that one and the same cause, a poison, can bring about symptoms apparently so different? The hallucinations, whilst maintaining the same general character, present degrees in their mode of expression which explain these differences.

Maniacal form. In a first degree, the patient thinks he hears insults or provocations, he sees robbers, armed men, animals; or perhaps he hears the voice of his relations or friends who call to him and warn him of danger or beg his help; or he sees his wife, surrounded by people, undergoing the most outrageous insults. Stung to the quick by these exciting causes, the patient answers, is insulting, quarrelsome, runs or rushes at his foes and becomes furious; all of these being acts which excite a noisy manifestation, a maniacal condition. This, as we shall see, is the form which is generally predominant in the attacks of *delirium tremens*, where hallucinations of a very lively nature are accompanied by trembling of the whole body, and a general tremor of the whole muscular system.

Melancholic form. In other cases the alcoholic sees himself in prison, before a tribunal, where he is accused of various crimes; he imagines that he has committed them, and thinks he is betrayed by all his friends; or he attends the funeral of his relations. Under the shock of these sad impressions he is gloomy, restless and suspicious, he bewails his condition, or he is frightened, tries to run away, and sometimes even conceives homicidal or suicidal ideas, and in a word, presents the aspect of a melancholic.

Stupid form. Lastly, in a still higher degree, he is about to undergo a terrible punishment; he has before him the bloody corpses of his children; everything is on fire; or he is going to be devoured, &c. These sights have astounded and terrified him, and he stays motionless in a state of complete stupor.

Between these different states, maniacal, melancholic, and stupid, we might insert numerous intermediate ones in order to reproduce faithfully the multiform aspects which alcoholic patients present; but as these divisions rest only on transient and passing characters, which have no effect on the course of the disease, and do not afford any therapeutic indication, they are of too secondary an importance to detain us.

ARTICLE II.

Mode of Evolution of the Sensory Disturbances.

I. *Sensory disturbances examined as a whole.*— Before we analyse these sensory disturbances, it will be useful to trace their development. We see, in fact, a successive gradation, both in the intensity of the phenomena and in their mode of evolution. They pass from a simple functional disturbance to an illusion, from this to an hallucination, at first confused and single, then multiple, and gradually becoming a definite, clear, and distinct hallucination, appearing, in a word, like the reality. In proportion as improvement occurs, the phenomena gradually disappear, following an analogous order in their decline, that is to say, the precise hallucination gives way to a confused one, this to an illusion, which in its turn is followed by simple functional disturbance. Such is the usual mode of evolution of the hallucinatory phenomena in alcoholism. In exceptional cases, however, they may at once attain their maximum.

It is usually at night that these symptoms first make their appearance, and although it is difficult to trace exactly their mode of evolution on account of the somewhat cloudy recollection of them which the patients retain, one may at any rate indicate with certainty their mode of disappearance. The hallucinations, which are at first persistent both day and night, begin to disappear in the day, whilst they still recur at night with the same intensity; then becoming less defined, at a later period they appear at that moment midway between sleep and waking which was long ago pointed out by M. Baillarger, as a period favourable to the outbreak of hallucinations.

Afterwards, we find only nightmares, which persist for some moments when the patient wakes, and then simply dreams, and the patient ends by appreciating with some degree of exactness those false sensory perceptions which soon entirely disappear. Thus, at first there are hallucinations night and day, and then only at night; later, there are confused hallucinations and illusions at the moment of transition from sleep to waking, and then during sleep, causing an uneasy awakening, and lastly, nightmares, dreams and return to health. It is easy to follow this declining course in most patients; as they get better they themselves observe with pleasure the favourable changes which take place, and some of them can recall them in a very striking manner.

II. *Sensory disturbances examined from the point of view of the peculiarities which they offer in each of the senses.*—Let us now trace the development of these phenomena in each of the senses. At the outset, the purely sensory disturbances resemble the symptoms observed in many other diseases.* As regards hearing, the first sensations are humming, ringing, and whistling noises in the ears; varied sounds, confused songs, the sound of bells, and then cries and tumultuous voices. Need we, however, with a German author, ascribe the ringings and hummings to irritation of the branch of

* Baillarger. *De l'influence de l'etat intermédiaére à la veille et au sommeil sur la production et la marche des hallucinations*, 1842. Baillarger. *Des hallucinations, des causes qui les produisent et des maladies quelles caractérisent.* (Mém. de l'Acad. de Méd., tom. xii. 1846). Brierre de Boismont. *Des hallucinations*, 1862.

the auditory nerve to the vestibule, the sounds and songs to the irritation of the branch to the cochlea, and the cries and confused voices to that of the two other branches?*

Be this as it may, the morbid interpretation of these sensations speedily comes into play, the sound of the bell becomes for the alcoholic a funeral toll; the cries and confused voices are insults and threats, wailings clearly expressed, or groanings, the entreaties of a parent or friend, or well known voices and clearly articulated words.

As regards vision, the symptoms occur in the same way; the sight is cloudy, becomes dim, and objects seem surrounded by a mist; then sparks appear, flames, varied colours, shadows, fluttering objects, grimacing faces, which enlarge and grow small, or approach and recede, then conflagrations, uproars and battles. In some cases the patient sees at first a dark blackish spot with confused outlines, then with distinct borders and with prolongations which become paws or a head, so as to form some animal, a rat, a cat, or a man.

The patient Anastasia D——, (Case iv.) saw on the wall intersecting lines and cobwebs; then in the midst of the meshes there appeared in succession, one, two, and three black balls, which enlarged and moved and took the form of cats and apes; they broke away from the wall, jumped on the bed, then ran away and went into the wall again; then she saw armed men, then women who threatened her, and finally the scene was completed and she was present at the murder of her own husband and children.

---

* Erhard. *Subjective auditory sensations.* (Med. Soc. of Berlin). Analysed in the *Arch. gén. de médicine.* March, 1868.

In patients affected with alcoholic delirium, in addition to the varied hallucinations of which we have just spoken, we may see amblyopia come on and last for a longer or shorter time; this has been previously noticed by many authors; in it objects become obscure, and seem to be enveloped in a white or greyish mist; the patients may confound gold and silver coins; they have trouble in sewing or writing, and sometimes they read with difficulty letters of three millimètres, or even larger.

With this amblyopia there often exists a disturbance in the perception of colours which M. Galezowski has endeavoured to elucidate, " By means of this method, viz. retinal chromatoscopy, we can" says the author, "discover from the commencement of the affection an anæsthesia of the retina as regards the perception of certain colours, and especially of composite colours which are neither clear nor deep. Such are especially the colours, yellowish-green and bluish-green, which the patients usually confound with the predominant tint. Violet is taken for red, and blue for grey. We often have occasion to observe at our clinique, patients suffering from alcoholic amblyopia, and up to the present time we have never found a single case in which the chromatic power of the eye had not been affected in some way or other. With improvement of sight, we have also seen the chromatic sensibility of the eye return."*

We have often enough found a transitory amblyopia during the attack of alcoholic delirium, but dyschromatoscopy is surely less frequent than M. Galezowski thinks.

The patient Louis C—— (Case xvii.) was a dyer,

* Galezowski. *On the diagnosis of diseases of the eye by retinal chromatoscopy.* p. 220, 1868.

and had consequently tolerably precise notions of colours, yet he took blue for violet, green for grey, and designated a red object by the name of chestnut or brownish mahogany (No. 10 of Dr. Galezowski's chromatic scale). This patient shewed, moreover, an enfeebled acuteness of vision, and could hardly read types of a millimètre and a half. (No. 6, Jæger).

The patient, Pierre L——, affected with right hemi-anæsthesia, could not read with the right eye types of less than 1 centimetre, whilst with the left he read fluently those of 3 millimètres. Moreover, with the right eye he distinguished red, green, and blue, but violet was white and yellow was whitish. The left eye, on the other hand, distinguished all the colours very well.

The perversions, illusions, and hallucinations, are less numerous and less varied in the senses of taste and smell ; but, nevertheless, the alcoholic does not escape either disagreeable odours or tastes; he sometimes smells odours of sulphur, of rats, and of putrefied matters; at other times, his food is sour or has an earthy taste, or seems like tainted flesh, or as if containing arsenic or vitriol.

In some cases they find in the broths given them, the taste of wine, brandy, or rum, whichever is their favourite drink. There is also occasionally, but much more frequently in chronic alcoholics, anæsthesia of smell and taste. For the exploration of the sense of smell, we employ substances generally known, with odours of gradually increasing intensity, such as orange flower water, camphor, essence of peppermint, tincture of musk, acetic acid, essence of mustard; we have thus a regular gradation, which allows us to measure to some extent the degree of anæsthesia. For the sense of taste, we also use in succession, sugar, salt, sulphate of soda, aloes, colocynth, and pimento; and it is well, in order to obtain an estimate of any value, either to use solutions, or to place the sapid body on the tongue and keep it there, preventing all displacement of the latter. This precaution is indispensable in cases of hemi-anæsthesia, as in the patients L——, M——, and D——.

The general sensibility with its different forms of anæsthesia and hyperasthesia also furnishes its quota of unpleasant sensations, and frequently its disturbances are associated with those of other senses. Thus some alcoholics feel and see animals crawling between the flesh and the skin; or they imagine that they are surrounded by iron threads which twine around, enclose and oppress them; they pass their time in unrolling these ever-renewing metallic circles; or they may even see a part of their body eaten up by worms which they shake, and try, in the liveliest anxiety, to detach and throw to the ground.

The patient D—— (Case iv.) felt some cold and damp creature creep along her thighs, and plunge its sting into her flesh; she felt movements in the legs, and the claws of an animal plunged into her back. The patient Louis D—— (Case xii.) felt the blade of a penknife cutting his throat.

The patient Pierre L——, (Case ii.) crushed a large black spider which was crawling between his trousers and skin. Balls were flung at another alcoholic which struck all parts of his body, dust which got into his eyes, and poisoned balls into his mouth.

It will suffice merely to mention certain anæs-

thesias which are more frequent in chronic alcoholics, limited to one limb or part of a limb, but
which may also affect an entire half of the body.
The breath, the touch, tickling, single or numerous
pricks, cold and warm bodies, electricity, &c., are
the means of exploration most convenient to use.
We shall return to this subject later when speaking
of the cases of hemi-anæsthesia with paralysis, in
which the special senses of the same side are also
involved.

ARTICLE III.

*Classification of patients affected with alcoholic deli-
rium in three groups, according to the clinical data
afforded by the antecedents of the patient and the
course and termination of the disease.*

I. *Patients affected with alcoholic delirium, with
easy, complete, and rapid convalescence.*—Having
studied alcoholic delirium as a whole, and expounded the general characters by which it may ·
be recognised, and having also analysed the numerous manifestations which it presents as regards each
of the senses, we must now follow it to its various
terminations. Sometimes, and indeed most frequently, the intellectual disturbance disappears at
the end of three or four days; and the patient begins to understand his own state, retaining, however, a certain mental indecision.

In the daytime he answers readily, speaking of
and laughing at his past visions and imaginary
ideas, but towards evening he loses some of his as-
surance and has vague fears, he feels already that he
will be less master of himself, and dreads the time of
going to bed; all this, because the illusions and

hallucinations will, as M. Baillarger has pointed out,
have just so much more sway as external causes of
excitement diminish, and as the will itself loses its
power and ceases to interfere at the approach of
sleep.

But at the end of some days long hours of peace-
ful sleep, undisturbed by dreams and nightmare,
restore quietude and strength, and the cure is rapid·
ly completed.

Cases of this kind whose termination is favourable are numerous and well known, two examples
will suffice.

CASE II. *Of drunken habits for several years—Abuse of wine and
sometimes of brandy—Attack of simple alcoholic delirium—Rapid
convalescence.*

Pierre M——, aged 42 years, pigeon-feeder at the market, has
been a soldier, and contracted the habit of drinking in Africa,
his present occupation also gives him frequent opportunities
for gratifying his taste for drink. Ordinarily he drank wine,
sometimes brandy and more rarely liqueurs. He had nightmare often, and slept badly; for a long time back he had
little appetite, spat and vomited sometimes in the morning white
or green phlegm. Having fallen ill in consequence of fresh excesses, he was carried to the Hôtel-Dieu, whence, as he disturbed
the rest of the other patients, he was sent to Sainte-Anne. On
admission, April 18th, he is incessantly moving; displaces everything about him, hunts in all the corners, looks behind the doors,
picks up from the ground imaginary objects which he shakes and
throws away immediately; strikes and rubs his foot upon the
ground as if to crush insects; passes his hand before his face and
blows to drive away threads and hairs, claps his hand on his
thigh, and seizing his trousers, presses hard to crush, as he says,
the great black spider which is creeping between the trousers and
the skin;—he looks through the window—"there," says he, "is the
band of La Place Maubert disguised as bears—there is a cavalcade
with lions and panthers, who are looking and making grimaces—
there are little children disguised as cats and dogs:" then he sees
two men threatening him, and crouches quite frightened; they
aim at him, he says, with their guns, they want to kill him be-

cause he has taken their daughter. He answers his comrades, he calls them, he hears disputes and wishes to run to them.

It is very difficult to arrest his attention; his hands and feet are incessantly on the move to seize or repel animals and objects of all sorts. His face is covered with perspiration, the skin is moderately warm; the temperature 38·2° C. (100·8° F.); pulse large and compressible, 80; tongue moist.

There is marked tremor of the hands, rather less of the legs, and quivering of the muscles of the face, especially when speaking.

Treatment. A dish of broth, lemonade mixed with wine, and two cups of decoction of couch-grass.

The patient was up nearly the whole night, talking, arguing, looking about on all sides, he saw policemen and armed men, and saw pigeons flying about the room, which he tried to catch; he pursued rats and cats; worn out with fatigue he lay down on the bed about two o'clock in the morning and slept two hours; then he rose and continued the previous day's mimicry.

April 19th. At the morning visit, he collects pieces of money on the ground, rubs the table; then placing his left hand on his knees which he presses together and raises slightly, he holds, says he, the pigeon, and tries with the right hand to make it swallow grain; he thinks he is at the market and accosts a number of people, afterwards he sees men dressed like savages, defiling on a rope. By speaking sharply one can arrest his attention, he has he says, headache, tingling all over the skin, and cramps in the legs.

He says that he is poisoned, that it is working in his body, the doctor at the Hôtel-Dieu has made him take, he says, the poison of La Pommerais, which he calls *vigitale* (*i.e.* digitalis).

The trembling of the hands is still considerable. Pulse large, 80. T. R. 37·8° C. (100° F.)

Treatment. Roast meat, a dish of soup; two bottles of lemonade, and a bath.

April 21st. Has slept part of the night, has seen pigeons in his room, and dead birds with long beaks, also rats, there were at least two hundred of them under the bed eating his provisions. In the day time he hears at times the voice of his comrades, sometimes also, threatenings or insult. The trembling of the hands is less. T. R. 37·6° C. (99·8° F.) Pulse regular, 76.

April 23rd. The hallucinations have almost entirely disappeared in the day time, but they still show themselves at night.

M—— talks freely about the hallucinations he has experienced, but for several days after this believes in the reality of some of them.

In this case we see the general features of alcoholic delirium well depicted. The hallucinations of M—— are multiform, changing, and for the most part painful, some recall his ordinary occupations; furthermore, they began to disappear first in the day time, afterwards at night, and at last there remained only nightmare.

The tremor of the hands and of the muscles of the face was very marked, but other bodily disturbances were relatively moderate; the absence of fever and of all complications permitted a favourable prognosis.

Liquids were given in abundance to favour the elimination of the poison by the urine and perspiration, more than this, the patient being left at liberty himself singularly aided the expulsion of the alcohol from his system by the lungs and skin, in the repeated movements required for the satisfaction of his delirious perceptions.

This case, so different from febrile delirium tremens, would by most authors have been classified as *delirium tremens,* and yet we could foretell a prompt and rapid cure. It is the confusion of all these cases, difficult sometimes to distinguish, which has given us such extraordinary statistics, in which we find on the one hand, a fearful mortality, and on the other, on the contrary, an astonishing benignity.

CASE III. *Of drunken habits—Two fits of alcoholic delirium with hallucinations especially related to the pre-occupations and profession of the patient.*

Pierre D——, æt. 33, a costermonger, has given way for a long time to habits of drunkenness, favoured by the nature of his business. For some time he has been irritable, has had rows with his mates, has lost his appetite and sleeps badly.

After several days of revelling, he was seized with delirium and was brought to St. Anne on July 15th, 1871. On admission he is restless and fidgety, sees all the hairs of his head falling off, opens his hands and shakes them to fling them to the ground; he sees and tries to keep off threads which float before his face. He hears the voice of his comrades, converses with his wife, speaks to her about the market, and the sale, then he suddenly stoops, and cowers in a corner to avoid, he says, the balls and the shower of stones which are thrown at him.

Perspiration profuse ; skin moderately warm ; pulse large, 90, easily compressed; tongue white : trembling of the whole body ; numerous contusions on the breast and arms.

Treatment. A bottle of Seidlitz water ; two jugs of barley water ; a bath.

For part of the night there were hallucinations and restlessness ; slept three hours.

July 16th. The hallucinations are less frequent, but still the patient is often on the move, speaks and wrangles with his mates. A bath ; two bottles of citric lemonade.

July 19th. The hallucinations have entirely disappeared in the day time, but occur at night, he sleeps but little, talks, cries now and then, and looks frightened.

The improvement continued on the following days, and D——— remembers the lively terrors which he has undergone, laughs at them readily, and rightly appreciating all these symptoms, promises not to drink any more and goes out of the asylum cured.

For four or five months D —— is moderate in his excesses, but by degrees returns to his former habits and is soon taken with a fresh attack of alcoholic delirium.

He re-entered Sainte-Anne on May 3rd, 1872.

On his arrival, he moves everything about him, undresses himself, hunts on the ground for pieces of money, pursues rats, and throws himself against the furniture ; then sees himself driven against a wall by a cart which is being backed, calls out to stop it, to go forward or he will be crushed ; he walks with the greatest caution, does not know where to place his feet, as he sees spread out on the ground around him cauliflowers, sorrel, radishes, &c., which he tries hard not to crush. Then he sees a horse's head, serpents, &c., or talks to his comrades, answers them, calls out, whistles, &c.

E

Trembling of the hands.

Treatment. A bath : two cups of citric lemonade.

The delirium persisted for two days with the same intensity, but on the second night the patient was calmer and was able to sleep for some hours.

Slight trembling of the hands.

At the end of a week the hallucinations had almost entirely disappeared in the day time, but still return at night with considerable intensity.

In the morning, D—— relates what he has seen and heard during the night, he shuts his eyes and stops his ears, he says, but it comes all the same, and then as soon as he begins to get drowsy, he is awoke with a start and requires some moments to get reassured. The improvement continued on the following days, and the patient is sent into the country.

This case gives us an instance of two attacks of simple alcoholic delirium, mild in character; the hallucinations, changeable and of a painful nature, revive most of the facts in the daily life of D——; in both fits they disappeared rather rapidly in the day-time, to appear again at night, especially at the moment in which the patient was just dropping off to sleep. Both attacks have been followed by a rather rapid amelioration, but if D—— does not forsake his drunken habits, fresh relapses will not, in all probability, terminate so favourably.

II. *Patients affected with alcoholic delirium, of slow convalescence and with ready relapses.* In this second group of persons, recovery is less rapid than in the alcoholized persons of whom we have just spoken; their sleep is troubled and unrefreshing; cerebral malaise, obtuseness and unsettledness of thought, irritability, excessive susceptibility to impressions, illusions, and vague ideas of persecution remain.

The patient, nevertheless, conducts himself with propriety, answers pretty exactly, but on pushing our questions in the direction of his occupation and persisting in this we soon come into collision with delirious ideas which at first were hardly manifest. This condition is found more especially in certain physiological states; we noticed it during the late Prussian siege in many people who thought they could with impunity replace an insufficient diet by a little more wine and brandy. We meet with this sort of thing rather frequently in some women who eating but little, drink, in the morning or between meals, either undiluted wine or spirituous drinks such as " Carmelites' balm-water" the "elixir of long life," the "vulnerary liqueur" or some other rubbish of this kind which they take " as a tonic" to give them strength and cure their stomach-aches. The following example will give us an idea of this group of patients.

CASE IV.—*Bad hygiene—Undiluted wine taken at breakfast— Alcoholic delirium—Hallucinatory disturbance of all the senses— Maniacal, melancholic, stupid forms in succession, according to the intensity of the hallucinations.*

Anastasia D——, æt. 45, a dressmaker, has for a long time been given to some excess in drinks, but more than this she has formed the bad habit of not eating in the morning and of taking either undiluted wine, or sometimes black currant wine. In December, 1870, and in January, 1871, she lived almost exclusively on bread soaked in pure wine. For a considerable time she has suffered from indigestion and often vomits in the morning, the ejections being white, or sometimes yellowish and greenish like bile. The loss of appetite, the indigestion and an intense thirst have more recently led her to drink and lately to take the Carmelites' balm-water.

The last three months she has slept badly: she is nervous; she hears the discharge of musketry, soldiers come into the house to kill her; she gets up, takes a light and looks on all sides, then is satisfied and returns to bed saying to her daughter, " how silly I am, I thought there was somebody."

The light being extinguished the hallucinations recur with the darkness; she tries at first to turn her attention to other objects,

2 E

she closes her eyes and strives to go to sleep, but in vain; suddenly she hears the voice of her relations, the groans and cries of her daughter who is being dragged away; she springs out of bed, tumbling violently against the furniture, runs to the window, and it is with great difficulty one succeeds in holding her back; the delirium persists until morning; she sees phantoms and birds.

Serpents glide over her bed; she sees fires; she hears a dreadful noise in the street where her parents are being massacred. She remains in a state of inexpressible anguish until daylight, when the hallucinations, without entirely disappearing, allow her some moments of repose. The following night the same phenomena are reproduced and the patient is sent to the asylum on April 2nd, 1872.

On admission she is a prey to the most varied hallucinations; every sense is the seat of hallucinatory disturbances, so numerous that we find in her so to speak the delirium of several alcoholics; she appears at one time maniacal, at another melancholic, at another stupid, these varied aspects occurring in a few hours.

As so varied a delirium is rarely observed, we shall classify the hallucinatory disturbances which we find jumbled up in our notes taken whilst following the delirium of the patient, according to the senses which were the seat of them.

Sight. She sees cobwebs on the wall; cords and threads in meshes which diminish and enlarge; amid these there appear black balls which swell out, contract, take the form of rats or of cats, these pass across the threads, leap upon the bed, then disappear; then she sees birds, forms making faces, and apes which run, come forward and go back again into the wall; on the window of the room she sees chickens which are escaping and which she tries to catch; on all the roofs of the adjoining houses appear men armed with muskets, through a hole in the wall she notices the barrel of a revolver pointed at her; she sees fires on all sides, the houses fall in and tumble to pieces; everything disappears; in the midst of this tumult she witnesses the massacre of her husband and children. A moment afterwards the trees seem to dance and are covered with globes of every colour, which recede, grow large, and lessen: at times immense fires variously coloured light up the horizon.

Hearing. She hears the voices of her daughter and her husband crying—fire—murder, and calling for help; she is insulted; she is called vulgar, a cow, &c.; she is threatened; she ought to be allowed to go, for they have already cut her infant up into morsels; they are going to burn everything; she hears bells, music, the noise of a machine on one side of the room; afterwards songs and confused cries.

Smell. She smells, she says, sulphur and vitriol, her clothes are poisoned; the counterpane smells rotten.

Taste. What is given her is sour and tainted; they wish to poison her, that has a taste of vitriol or something that she cannot tell, but it is poison.

General sensibility. She feels a pricking over her stomach; something heavy pressing on the skin; a cold and wet nasty thing crawls over her thighs, it drives its sting into her flesh, she feels movements in her legs, and the claws of an animal fasten on her back.

Maniacal form. The patient very restless, covered with perspiration, incessantly on the move, passing her hands over her body, over the walls, and moving them in the air, pursues imaginary beings, cries, answers, calls; wishes to open the doors, hides herself, defends herself, &c.; in fact is in a most typical maniacal condition.

Pulse large, soft, regular, 80; tongue white, epigastrium tender on pressure. Trembling of the arms and legs.

Treatment. A bottle of Rogé's lemonade, two jugs of citric lemonade.

Melancholic form. The hallucinatory disturbances continue all the night; but the patient being very frightened, and in a state of great anxiety, weeps and groans.

Stupid form. For some hours she does not stir a foot, answers no question, remains motionless, allows herself to be moved, makes no effort to get rid of the saliva which fills her mouth and runs over her lips, she passes her water under her, and remains in a condition of complete stupor. The next day, in the day-time, the patient is uneasy and restless. She sleeps some hours the night following.

April 4th. D—— is sad, she still hears the cries and groans of her relatives; she has seen her daughter with her arms cut: she is insulted; they wish to kill her. Heaviness of head; trembling of the hands; cramps in the legs. The patient sleeps part of the night; she is less hallucinated; but the hallucinations leave behind them ideas of persecution; she is in prison; they wish to get rid of her; to poison her; she remains uneasy and pre-occupied, and distrusts those about her.

Treatment. Two pots of lemonade with wine ; a dish of soup ; a bath.

April 6th. Improvement continues, but the patient remains sad, and retains the idea of persecution. Heaviness of the head ; numbness of the limbs ; the trembling of the hands is less. Patient begins to eat.

Treatment. Two bottles of vinous lemonade ; a dish of soup ; a bath.

Fifteen days later, D—— still retains some idea of persecution, she is sent to another asylum.

This case gives us a sufficiently clear idea of alcoholic delirium and of hallucinatory disturbances. These hallucinations are almost all painful and mobile, and recall most of the persons or things which are habitually associated with the patient ; they recall also some of the scenes of which D—— has been witness during the late events, and by which she had been particularly impressed according to the account of her friends, and they, moreover, ascribed the delirium to the emotions which she experienced during the Commune. The hallucinations, as they have disappeared, have not been followed by a complete restoration to health, there still remain some ideas of persecution, which undoubtedly will remain for some time yet, and furthermore the headache and numbness of the limbs show that the intoxication has already made some progress.

In a certain number of alcoholics, after the acute symptoms, we find delirious conceptions which arise from existing or past hallucinatory disturbances persisting, sometimes multiple and variable, sometimes more limited and assuming the form of a partial delusion. Thus it is that some who are jealous and suspicious, think themselves deceived by their wives and interpret in this light everything which is said and done around them. Others, exposed to prosecutions, hear themselves accused of obscene acts; complain of attacks on their honour and on their morality; people wish to ruin them, to plunge them in misery, to get them out of the way. Sometimes, again, there remain disturbances of the general sensibility with hypochondriacal ideas, and fears of poisoning.

We give in this place two examples of this kind; we shall find further on cases belonging to this category, certain facts in whose history, such for example as the abuse of absinthe, have led to their being placed in one of the following articles.

CASE V. *Admitted six times into asylums with alcoholic delirium, from the year 1855 to 1870—Two slow recoveries.*

Frank D——, æt. 39 years, a strolling singer, whose father appears to have died insane, has been under treatment four times at the Bicêtre; the first three times in 1855 and 1856, the fourth in 1858.

Vexation and excess in drinking, he states, have always been the cause of his illness. In his first attack he threw himself into the Seine, from the top of the Pont au Change; taking the parapet for a person blocking up his way, he freed himself by a bound and fell into the water.

Since 1855, after an attack of mania which lasted six months, he has given up his profession of an architect's clerk, not being able, he says, to bear the work of the office, and has become a strolling singer.

His new kind of life has led him into all sorts of excess. Six weeks before his last sequestration he began to feel the marked symptoms on account of which it was necessary to isolate him.

Every night he dared not blow out his candle for fear of witnessing again the fantastic visions of the day before, and the hallucinations of hearing which followed him everywhere. Very distinct voices repeated his thoughts, insulted him, and wearied him with their tiresome discourse. As soon as he was in the dark he saw all sorts of objects assuming frightful dimensions. If he saw a basket of flowers on his bed, it became of immense size, and peopled with thousands of disgusting animals, serpents and toads; he was perfectly conscious of these hallucinations, but was none the less beset on that account. If he closed his eyes, he was roused from his sleep by nightmares which woke him in starts.

To overcome the fatigue brought about by want of sleep, and to sing as usual, he drank a good deal and was often drunk.

On the day of his arrest while passing over the Pont Neuf, he heard voices threatening to kick him if he did not throw his clothes over the parapet. He was convinced, he said, of being at this moment out of his mind, but he was not master of himself; and he was taken to the police station, when he was about to take off his clothes—talking meanwhile in a loud voice.

On admission to St. Anne, on May 8, 1869, he is agitated, frightened, thinks he is pursued, looks on all sides, answers individuals who are insulting and accusing him, protests his innocence.

The skin is covered with perspiration; the pulse large, frequent. and compressible; the tongue white at the base; thirst extreme. The urine of a dark yellow colour, is not affected by heat, it gives as light effervescence with nitric acid and is unaffected by Bareswill liquid. Trembling of the hands.

Treatment. A basin of barley water with 10 grammes of cream of tartar; two basins of couch-grass tea.

May 9th. Has had a bad night; hallucinations incessant; the patient was out of bed several hours talking, threatening, and calling for help.

This morning he is worn out, less subject to hallucinations; his attention is more easily directed and he answers questions pretty exactly.

The tongue is moister; bowels relieved.

Treatment. A bath; three bottles of citric lemonade.

May 12th. Sleeps during a part of the night; there are some hallucinations, but the patient is quiet, keeps in bed and does not talk.

In the daytime the sensory disturbances are rare. D—— begins to give an account of his situation, and to doubt the reality of what he saw and heard during his delirium. From this time the improvement continues, the patient finds something to do, behaves well, sleeps tranquilly, and after a stay of three months departs, promising not to drink any more.

D—— is scarcely at liberty, before, having resumed his calling as a strolling musician, he again falls into excess, drinking brandy and absinthe. Headache, dizziness, vertigo, and hallucinations, are not slow to shew themselves.

One evening, whilst he was playing in a café, he imagined that everybody was making fun of him, insulting and threatening him. He suddenly left the room and fled away quite scared, was arrested in the street, and 48 hours afterwards was brought to the asylum, November 20, 1869.

He hears threatening words, sees rats around him, and spectres which enlarge and successively advance and retire. His sleep is restless, nevertheless the patient keeps in bed.

In the morning he suffers from headache and lumbago; his hands are slightly tremulous. The urine is free from albumen and sugar.

Treatment. A bath; a basin of gentian tea.

December. The sensory disturbances have completely disappeared in the daytime, but the patient remains gloomy, is ashamed of his conduct, deplores all that has occurred; in the night, dreams and nightmares still disturb his sleep; sometimes he even sees phantoms. From time to time, he suffers from headache with dizziness.

March. Restoration progresses; the patient works, regrets his excesses, promises to drink no more, and appears decided on changing his occupation. He leaves at the beginning of April.

Notwithstanding his hereditary antecedents D—— has improved rather rapidly after each relapse, especially in the last two attacks; but still, in 1855 the alcoholic symptoms have been followed by a fit of mania which lasted six months; and on his admission the last time some melancholic ideas remained at the end of three months, and also headache and giddiness, which show plainly that we had not to deal with an alcoholic delirium of mild, rapid and complete convalescence. Consequently, if fresh excesses should again bring D—— to the asylum we should fear that a variable delirium might persist, and that its duration might be very prolonged; or perhaps a weakening of the faculties, which as yet has only reached a slight degree.

On leaving the last time D—— promised and seemed resolved to change his business, since that date he has not returned, perhaps he has kept his word and is cured, but the events of 1870 and 1871 have caused the disappearance of so great a number of these unfortunate patients that we must put a query here.

CASE VI. *Typhoid fever when 22 years old—Excess in drink—Alcoholic delirium followed in its first attack by a convalescence interrupted by transient fits of mania—Three other attacks of delirium taking a more regular course.*

Clement L—— is a man 33 years of age, a costermonger by trade. His constitution is apparently quite robust, the muscular system well developed, with a tendency to obesity; his neck is short, and his face colours deeply under the slightest influences.

When about 22 years old he had typhoid fever; from that time his character has undergone a change, and his mind has remained somewhat blunted; at times he will be excited and constantly on the move.

For some time past he has abandoned himself to drink, and has often been intoxicated, he drinks more especially white wine and brandy, and absinthe occasionally. Now and then he suffers from headaches, dizziness, buzzing in the ears, and muscular weakness, but no true paralysis; whilst under the influence of recent excesses repeated for several days, he was seized with alcoholic delirium, and came into St. Anne, on September 15th, 1868.

We noticed in him marked restlessness, disorder in his ideas and actions, a propensity to finger everything he sees, periods of passion and violence, tremor of the tongue and of the hands. At times he thinks he is surrounded by persons who threaten him and seek to strike him and to kill him.

On examining his urine neither sugar nor albumen is found; it is thick, becomes clear on heating, and effervesces when treated with nitric acid.

Treatment. A bath; a basin of barley water with 10 grains of cream of tartar; two dishes of couch-grass tea.

Sept. 18th. The restlessness is much abated, his sleep has been quite calm, and the hallucinations less distinct.

On the 20th, the day of his removal to the Asile de Ville Eyrard,

L—— is quiet, his behaviour is more correct; his answers quite clear; he gives a closer attention to questions that are put to him; his intellectual faculties do not appear to be notably weakened.

At the visit at the new asylum on the 21st; the improvement has not been maintained; on the contrary the restlessness has returned very violently; the terrifying hallucinations have reappeared. As the face is red and swollen, bleeding is prescribed; and the symptoms subside rather rapidly.

At the beginning of December, for the purpose of hastening his cure, the patient is allowed to go and work in the garden; his condition appeared quite satisfactory, when all at once, he is seized with delirium, becomes violent again, incoherent and coarse; four scarifyings at the nucha are ordered. This sudden disturbance is gradually succeeded by a perfect calm.

Another crisis soon reproduces completely the same physical and intellectual disturbances.

Since the beginning of January the calm has not been interrupted, the delirium and the other morbid phenomena have entirely vanished. (The notes taken during L——'s stay at Villo Evrard we owe to the kindness of Dr. Drouet).

Permission to leave the asylum was given March 4th, 1869.

L—— came again into St. Anne, on October 28th, 1869.

According to his brother's account he has never given up drinking to excess since he went out; about a fortnight ago, after more copious libations, (wine and brandy), he was once more delirious and excited, and was taken into custody in the suburbs of Paris. The disturbance of the intellect is intense, the incoherence of his talk very great, with marked excitement; he hears voices calling him from all sides, he tries to pick up objects which he thinks he sees on the floor, he sees lights, flames and sparks; his face is injected; there is trembling of the tongue and hands.

The patient goes to another asylum.

August 30th, 1870. The third sequestration. L—— having passed the whole day in drinking is seized with violent agitation.

On admission, the attack has lessened in intensity, there remains disorder of the ideas and actions, incoherence in language, slight excitement at times and some hallucinations.

This condition remained the same up to the time of his removal to another asylum.

On June 21st, 1870, having been picked up drunk in the public thoroughfare, L—— on his arrival retained no recollection of

what he had done. He is slightly excited, imagines himself surrounded by apes which are making grimaces, by animals of various kinds, and by armed men who are talking of shooting him;
he spits up some phlegm in the morning, there is a little trembling
of the hands, and he has cramps in the calves of his legs.

June 23rd.   The night has been a restless one, he got up
several times and appeared to be still subject to his hallucinations.

On awaking he is talkative, incoherent, prolix in his answers,
and unable to tell what he has experienced.   When removed to
another asylum, improvement became more evident day by day,
there remains only a little headache, and sometimes, in the evening he complains of visual disturbances such as flashes of light,
and dazzling sensations; his face is at this moment more congested, the trembling of the tongue is slowly passing away.

Every day L—— goes into the garden and does his work exactly, without any word or action that attracts attention.

His improvement continued during July and August, he then
left.

A prior illness from typhoid fever had left some
persistent impressions on L——s character and
intelligence which had already been noticed by his
family. Under the influence of repeated abuse of
alcohol we see fits of delirium making their first
appearance, associated with hallucinations of varying intensity which got rapidly better by his re-
siding in the asylum.

During his first sequestration we noticed the fact,
which occurs sometimes, of the sudden onset of two
attacks of excitement after periods of complete calm,
and that without fresh libations. These accidents
are accompanied with dizziness, buzzing in the
ears and congestion of the face; from L——'s
complexion he appears to be predisposed to symptoms of this kind; in the course of the last three
attacks of delirium, although the influence of the
alcohol has gone on increasing, we see nothing
similar: we know that, in alcoholics, phenomena

like those they experience after recent excesses, may arise quite suddenly from various causes. Furthermore, we must not forget that at the time L—— suffered from these cerebral disturbances accompanied with excitement, he was in a locality where intermittent fever is common, and that he might possibly have shewn the exacerbations of which we have spoken under the influence of some slight attack of malarial origin.

The delirium of the form of depression, manifested by some of these alcoholics, which is limited in certain cases, allows, like partial delusions, a certain freedom of mental action to the patients, who, being thus able to pass more easily from delirious conceptions to actions, furnish the most numerous examples of homicide and suicide.

All authors, and with good reason, mention drunkenness as one of the principal causes of suicide. One fourth of the suicides in Berlin in a period of nine years, according to Casper's statements founded on official documents, comprised people given to drink. M. Brierre de Boismont, in his work on suicide and suicidal insanity, gives 530 cases of suicide through drunkenness in a total of 4595, or almost one-ninth. The following are figures which we borrow from the general statistics of the insane admitted into the asylums of the department of the Seine in 1870 and 1871, which M. Bouchereau and myself communicated to the Medico-Psychological Society.* In 1870, among 1460 insane people, 377 were alcoholics and of these 28

* Bouchereau and Magnan, *Statistique des malades entrés en* 1870 *et* 1871 *au bureau d'admission des aliénés de la Seine.* (Ann. Med. Psych., 5th series, vol: viii., Nov. 1872).

had made attempts at suicide, and 9 had attempted homicide.

Out of 1059 insane, 64 were alcoholics and of these 9 had made attempts at suicide.

In 1871, among 1128 insane, 291 were alcoholics, and of these 24 had attempted suicide, and 8, homicide. Out of 1070 insane, 64 were alcoholics and of these 10 had tried to commit suicide, and 1, homicide.

Our figures are less than those given by other authorities for two reasons : we have cut off the list of homicides and suicides all accidents (and they are numerous) arising in a casual manner. For instance, a person affected in this way by alcohol imagines he is pursued and jumps through a window thinking it is the door, this is an accident and not an attempt at suicide ; another sees in front of him a phantom, an armed man, he seizes a chair, strikes and injures his child which is lying near him; this, again, is an accident and not an attempt at homicide. Furthermore, we have only mentioned cases on which we possess positive information. This statistical result should therefore be considered a minimum.

Delirium of expansive form.—Ideas of an exalted character, gay or ambitious, occur but rarely in the first attacks of alcoholic delirium ; we have had occasion, however, to observe some instances of it ; but together with the ambitious ideas, we find also the usual delirium of painful character. Thus, an inspector of police admitted into St. Anne in April 1869 for an attack of simple alcoholic delirium of which he had got well at the end of three weeks, had had during the first days a slight attack of delirium of an ambitious character mixed up with painful hallucinations. He thought he was a millionaire, that he ought to live in fine style, to have mistresses, servants, and carriages, and that everybody about him ought to be happy. Some moments afterwards he saw robbers coming to take his fortune from him, assassins armed with knives killing his father and mother, he noticed blood around him, spectres, and ferocious beasts; they called him the murderer of Montmartre, etc. Later on, in his convalescence, he perfectly remembered this delirium and said, "it was a mirage, but it appeared quite real to me." In chronic alcoholism ambitious delirium is of rather more importance, since, in some doubtful cases, it may aid us in recognizing the tendency of chronic alcoholism towards general paralysis.

III. *Patients specially predisposed, who when suffering from alcoholic delirium, have frequent relapses and a convalescence often interrupted by delirious ideas, assuming more or less the form of partial delusions.*

Persistence of delirium, after the acute symptoms, reveals itself in alcoholics of special predisposition, in those predisposed persons of whom M. Moreau speaks in his *Psychologie Morbide*, and to whom M. Laborde has drawn attention in a recent work.*

The cerebral inferiority of these persons finds its most frequent cause in heredity. But there is also

---

* Moreau (de Tours), *La psychologie morbide dans ses rapports avec la philosophie de l'histoire*, 1859.—Laborde, *Les hommes et les actes de l'insurrection devant la psychologie morbide*, 1872.—Morel, *Traité des maladies mentales.* Paris, 1860, p. 513.—Morel, *Traité des dégénérescences physiques, intellectuelles et morales de l'espèce humaine.* Paris, 1859.

the group of individuals who, thanks to repeated excesses, create for themselves, so to speak, a morbid predisposition. It is interesting, from this point of view, to compare the condition of their intelligence at the various dates of their admission into asylums; on each relapse the individual is gradually transformed, he descends in succession the various degrees of the intellectual scale, and in the third or fourth relapse we very often find weakening of the faculties behind the alcoholic delirium.

In this class of patients alcoholic drinks act in a manner different from that on other subjects; one is surprised to find on each relapse so marked a delirium accompanying motorial disturbances of but slight degree, there is not the correlation between the physical and the intellectual symptoms which we find in the generality of cases. In these alcoholized persons the nervous system seems to divide itself, all that appertains to the intellectual order shews an excessive susceptibility to impressions, whilst all that is concerned with the physical order carries on its functions in the normal fashion, there is a loss of equilibrium; we judge rightly about this want of equilibrium by observing the phenomena which occur under the influence of one and the same agent. Alcohol, for example, in animals as in man, gives rise to physical and to intellectual phenomena which ordinarily advance together, develope, and increase in parallel order, and this consensus of symptoms is in accordance with the action of the poison, as spread over the whole cerebro-spinal system; but here the equilibrium is broken, a weak dose of poison suffices to make a lively impression on the brain in its intellectual part, whilst on the contrary, physical disturbances, symptoms belonging to the spinal cord, or to the nervous system of organic life, are developed but little, or at least do not present a degree of intensity beyond what one usually observes.

Moreover, what occurs in this case as regards intellectual phenomena, is produced in the same manner in other diseases; it is the fact of the existence of a weak organ in certain individuals, which is always liable to give way under the most varied conditions.

CASE VII.   *Father a drunkard—bad instincts—alcoholic excess early in life.   Admitted 9 times into the asylum from 1864 to 1872. Numerous misdemeanours.*

Desiré L——, æt. 28 years, born of a drunken father and tuberculous mother, went to school until his 10th or 12th year, and learned quite easily to read and write.   From his 12th to his 20th year he had many a reckoning to settle with the police for various misdemeanours (stealing fruit from gardens, vagrancy, fighting, &c.), for all such things, as he said, as *gamins* are in the habit of doing.   Even so early as this he began to drink.   He slept badly and frequently had nightmare; and sometimes headache.

When 20 years old he had typhoid fever, which was marked by no particular symptoms.

Excess in drinking became habitual to him; his nights were bad; he became quarrelsome, irritable, and could not bear the least contradiction.

In his 21st year he had the first attack of alcoholic delirium, he absconded from home and ran about the streets thinking he was pursued by assassins.   He was taken to Bicêtre and went out cured at the end of four months, (1864).

He was scarcely at liberty before he resumed his old style of life, and was not long before he had fresh symptoms which brought him again to Bicêtre, whence he was sent to the asylum of St. Albans with a convoy of sick pensioners of the Seine.   Sent away from St. Albans as cured he returned to Paris.   He continued to drink, he soon imagined that people called him names; he saw people placed at the street corners to watch him; persons whispered in his ears.   He rambled about in all directions, living

F

in a wretched fashion ; was arrested for vagrancy and sent to St.
Anne, December 10th, 1867.

On admission he is sad, and sorrowful, and maintains that persons bear him ill-will and make him wretched, and that they are
always after him.   Now and then he sees sparks and sudden
flashes before his eyes.

Heaviness of the head, dizziness ; trembling of the hands.

Treatment.   A bath every second day.

December 15th.   He is quiet, answers questions willingly; no
longer experiences hallucinations in the day-time; but he is sad,
appears to be pre-occupied; he sleeps badly, awaking frequently
with a start.

Headache; slight trembling of the hands.

Treatment.   A mustard foot-bath in the evening.

January.   The patient remains in a stationary condition ; at
times he is irritable, appears unsettled, pre-occupied, seems dissatisfied with the interest and attention bestowed upon him.   In
the night his sleep is often disturbed by nightmare.

To promote convalescence L—— is sent to an agricultural
asylum.

He spent six months in the country, attending with regularity
to easy work, but he was reserved and wanting in frankness, and
shewed from time to time an excessive impressionableness.

He escaped from the asylum and returned to Paris.   He began
drinking again and went to the prefecture of police to demand his
papers and to protest with the greatest earnestness against the
arbitrary confinements of which, he said, he was the subject.

Having been brought again to the asylum on Aug. 24th, 1868,
he was again sent into the country.   There he worked in an orderly
manner, and behaved himself well; his sleep was calm ; L——
appeared to be in a good way for recovery.   He was allowed to
leave on trial on October 15th, 1868.

On his return to Paris he went to see his old friends, walked
about, visited several quarters, most probably he committed some
excess, and drew up more complaints against his sequestration.

He wrote to the prefect of police, went to find the commissary,
was angry, flew into a passion, gave full vent to incoherent notions about persecution, and came into St. Anne for the fifth time on
Oct. 21st, 1868.   In a few days the passing excitement was sub-
dued ; L—— resumed his habitual character, worked in the house,
appeared full of good resolutions and obtained permission to leave
at the beginning of June, 1869.

He was scarcely out of doors before he was drinking again, he could not apply himself to any work, and was drunk one day in a tavern and struck two customers sitting at the table by his side. For this act he was sent to Mazas prison and soon brought back to the asylum, which he entered for the sixth time on July 15th, 1869. On admission he was almost well of his late attack; he related that several days after he left he noticed that people looked cross at him, pointed their fingers at him, made fun of him, and that three persons had insulted him in the wine-shop; he did not know them, but was quite sure that they had spoken ill of him. At night he had no sleep; he saw flames and glimmers, he heard a noise in the distance, sometimes cries and confused voices. He ate little and drank much.

After a few days he was sent to an agricultural asylum where he remained until December 1869. Being set at liberty, he spent nearly a month in Paris, and was brought back again for the seventh time in an attack of alcoholic delirium, on January 27th, 1870. At night, he said, some one wakes him up, calls him, enters his room; he is obliged to get up, and sees nobody; when in the streets, people look black at him, hinder him from working, and place obstacles in the way of everything he should do. He relates how, the other day, being in a crowd, he saw persons who wished to do him harm; he fell upon them and struck them, (this in fact it was which caused his being sent into the asylum). He has no appetite; his tongue is white, his hands trembling.

Treatment. A bottle of Seidlitz water; two tumblers of citric lemonade.

At the end of a few days the sub-acute symptoms subsided, and L—— was sent to the country asylum, whence he came out in the month of August 1870, at the time of the war.

Peace being restored in Paris after the painful events of 1870 and 1871, the delirious acts of L—— which had passed unnoticed during this unsettled period, were not slow to attract attention, and to cause on the 12th of June, 1871, his eighth confinement. This time, after recent excesses in drinking, he had struck an innocent passer by on the boulevard, under the idea that he was following him as a spy.

On admission, L—— presented hallucinations of hearing and of sight, he heard disagreeable things people were whispering about him; he saw flames and clouds which rose and fell before him.

Headache; trembling of the hands. L—— having returned to

the quiet and regular life of the asylum got better quite rapidly, and on the 18th of July, 1871, was transferred to the asylum of Pont-Labbé, whence he was set at liberty at the beginning of 1872.

On his return to Paris, he again struck two persons in a wine-seller's house, was sent to Mazas and came into the asylum again for the ninth time, on April 15th, 1872.

"People had begun the same plagues again," he said, "they made a row about my room; they hustled me on the staircase; they abused me." At times, he saw figures, heads, and trees, which disappeared again suddenly.

Headache and frequent giddiness, sometimes pains along the course of the nerves of the face, more particularly the supra- and infra-orbital.

Tongue white; tenderness at the epigastrium on pressure; thirst; anorexia; no fever; nothing particular in the urine.

Treatment. A bottle of Rogé lemonade; two basins of barley-water.

June. Improvement occurred more slowly this time, ideas of persecution persisted, with hallucinations at night and sometimes in the daytime. L—— was sad, "fate," said he, "pursues me relentlessly, I am always unfortunate;" he often had headache, and sometimes numbness in the limbs.

From the 20th to the 28th of June, he had gastric derange-ment with slight fever, and during these few days' malaise the hallucinations became more frequent; at night he saw persons pass before him, he heard the noise of confused voices around his bed; in the day-time, even, the visual hallucinations were repro-duced if he shut his eyes.

July. He is a little better, he occupies himself part of the day, he sleeps more peaceably; but he often has headache and giddiness; he still has trembling of the hands, and occasionally numbness and cramps in the limbs.

The history of this unhappy patient, whose ex-cesses in drink, favoured by a grievous predisposi-tion, constantly placed him in contest with society, is a real odyssey. From his youth he showed un-healthy tendencies, and we see alcohol acting on this easily impressed organism with an energy quite peculiar. Some slight excesses have repeatedly been sufficient to bring back again the alcoholic symptoms and hallucinatory disturbances, after which, there persist melancholic ideas and notions of persecution, each time of a more marked character.

Equally worthy of notice is the momentary appearance of hallucinatory disturbances under the influence of a slight intercurrent affection, (from June 20th to 28th, 1872,) during his stay in the asylum, in the absence, consequently, of any fresh excess.

A transient disturbance of the physiological equilibrium sufficed to allow the intoxication, which had become latent, to manifest its existence in a sudden and noisy manner. In the same way, again, attacks of alcoholic delirium appear, owing to some intercurrent affection, in drunkards who have not committed any excess for some time past. The system, which in health is strong enough to resist the intoxication, finds itself defenceless when the disease has weakened it and disturbed the co-ordination of its functions.

The prognosis is certainly very grave, whether looked at from the point of view of physical or of mental disturbance; in both cases the limit of a possible cure seems to have been over-stepped.

CASE VIII. *Hereditary antecedents; father committed suicide; brothers of nervous temperament; a child died of convulsions.—Habits of drunkenness; hallucinations with ideas of persecution; attempt at murder; first admission to the Asylum, inmate for seven months; improvement.—Fresh excesses—Relapse—Second admission to Asylum. —Alcoholic symptoms which improve, leaving behind them hallucinations and ideas of persecution.*

Jean J——, æt. 27, formerly a soldier, was born of a melancholic father who was addicted to drinking and poisoned himself; he has had six brothers, the eldest, a drunkard, has led a disorderly life and squandered the possessions of the family; the youngest who is now 23 years of age had convulsions in infancy, is weak in mind and affected with strabismus; the third is in good health; the fourth died of an intercurrent disease, and the other two of convulsions. He himself has had two children by one woman, one at full term was born dead, the other died when seven months old after attacks of rigidity and shaking in the limbs.

From infancy he has had dreams and nightmare during the night; he has always been gloomy, sad, and sensitive, and early in life he formed the habit of drinking. In 1870, after more frequent excesses, he lost his sleep, was disturbed by hallucinations, he imagined that people bore him ill-will, insulted him, sought to ridicule him, that his mistress deceived him and schemed with other persons to get him killed. Under the impulse of delirious ideas he attempted to murder this woman by striking her with a knife in the street. He allowed himself to be taken without any resistance, spent six weeks at Mazas, and after a medico-legal enquiry was sent to the St. Anne's asylum, August 15th, 1870.

He slept badly, saw things in the night that frightened him; heard men about his bed who wished to kill him; heard the voice of his relatives; people reproached him. He had headache, loss of appetite, and slight trembling of the hands.

Treatment. A bath every second day; a dish of gentian tea every day.

At the end of two months the hallucinations had diminished in the day-time; but the patient was sad, restless, and wanted to be alone; his nights were bad and disturbed by frights and nightmare. He was transferred to another asylum, which he left in March 1871, improved, but not entirely cured. He was sad, irritable, and his sleep was seldom calm. Once at liberty, he was not long before resuming his drunken habits; he drank wine, vermouth, and brandy; he spat in the morning, and sometimes vomited sour fluid. He was re-admitted into the army on July 8th, 1871, where he remained until the end of his service, February 26th, 1872.

On his return to Paris he resumed his old trade as a hair-cutter, and during this time continued his excessive drinking. He became restless; at the shop he lived alone, retired, believing himself exposed to the puns and raillery of his comrades; persons came into his house, followed and watched him. To get away from his enemies he left, went into the country; heard cascades roaring around him; saw an illuminated castle, panoramas; he heard songs, music, and saw red lanterns swinging, suspended from the trees. He walked quickly, fled; all these phantasmagoria, he said, followed him; he saw crosses, shadows, sparks; if he went forward, everything got farther off; if he stepped back it approached; at times all disappeared. He heard, he said, voices which urged him to destroy himself; "the noise of the water and of the cascades was to defy me; it meant that I was a coward and that I had not the courage to throw myself into the water." "In the cottage where they played music, and danced and laughed, I understood it well, it was a lot of people who desired my ruin, mocked me, and sought to assassinate me."

"I wanted to drown myself, I went on the Pont de la Marne, men were there and I could not cast myself into the water; misfortune pursues me; I wish to do well and yet do nothing but wrong; I have deserved it all."

After having presented himself three times at the house of the Commissary of police, for arrest, he was brought back to the asylum, on April 10th, 1872, with this certificate from M. Lasègue; "Subacute alcoholism, probably hallucinations; fears, accuses himself of imaginary crimes, second attack."

On admission he appears restless, pre-occupied, remains alone; his look is brilliant, his eyes wander in every direction, at first he refuses to answer and seems to mistrust those around him.

During the night, he gets up, walks about his room, speaks for a moment in a low voice, or else cries out and threatens.

11th. Morning. He is calmer, more trusting, the hallucinations are less frequent; he relates pretty exactly what has happened since his first admission.

During the night he has seen, he says, five persons watching him and calling him an assassin, who wished to do him injury; he heard flies buzzing about his ears; he saw lights fluttering about; he heard the voices of his relatives and of his comrades reproaching him.

Heaviness of the head; skin warm, moist; pulse 80, large; tongue white; anorexia.

Trembling of the hands; cramps in the legs.

Treatment. A bottle of Seidlitz water, two bottles of citric lemonade.

13th. J—— is sad, he hears threats and insults especially at night; persons walk into his room; he sees spectres. In the day-time he gets away from the other patients and avoids speaking to them.

Treatment. A bath; two bottles of citric lemonade.

April 19th. J—— answers more willingly, his countenance is

more open, but he spends bad nights, disturbed by hallucinations.
In the daytime he still hears disagreeable words occasionally;
he believes in the reality of everything he has seen and heard.
"You won't make me believe that it was not true; the physicians have been bribed against me." Headache and heaviness of
the head at times; cramps in the limbs; persistent trembling of
the hands.

May. He is calmer, but often just as he is going to sleep he
hears tumultuous voices in the distance threatening him; he sees
jets of flame flit before him like shooting stars. Sometimes he is
no sooner asleep than he wakes up in a start, quite frightened, it
seems to him that he is just about to be shot, it takes a considerable time to remove his fears and his sleep continues disturbed all
night.

In the daytime he occasionally thinks that people are mocking
him, and alluding by words of double meaning to everything that
has happened to him.

He remains gloomy and reticent, becomes irritable and is ready
to fly into a passion on the least opposition.

The frontal headache persists, accompanied by hissing and humming noises in the ears and by numbness in the limbs.

June—July. There is but little change; the same pre-occupa-
tion of mind; ideas of persecution and delirious interpretation of
whatever occurs about him; illusions and sometimes hallucinations in whose reality he thoroughly believes; "I shall get rid of
these notions," says he, "when I am dead." His sleep is slightly
disturbed by nightmare. He still occasionally sees flames and
sparks, and hears threatening voices. At times he has headache,
humming noises in the ears, and cramp in the calves of the legs.
The hands still shew some degree of trembling.

Hereditary influence made itself felt in the case
of J—— in the most evident manner; he was a
predisposed patient in whom the toxic action of
alcohol shewed itself in two ways:

1st. By producing the peculiar delirium of alcohol.

2nd. By acting as an excitant on a prepared
ground and by thus favouring the outburst of a
delirium which might not have been produced in

the absence of some stimulus. We see how slowly the hallucinations, even the characteristic ones, disappeared; we find still in June and July hallucinatory disturbances which remind us of the toxic origin of these symptoms; but what is more important to notice is the deep conviction of the reality of all these illusory perceptions. Just as simple alcoholics (*i.e.* those without hereditary tendencies) aid the efforts of the doctor as soon as improvement begins to shew itself, by being easily convinced of, talking about, and imperceptibly appreciating, the true meaning of their sensorial derangements; so on the other hand, these alcoholics with hereditary predisposition are rebellious to all treatment, physical or moral.

ARTICLE IV.

Action of absinthe on man.—Muscular shocks; Vertigo.—Epileptic attacks; want of relation between the epilepsy of absinthe and trembling.—Hallucinations; Delirium.

In the physiological part of our work we have studied the action of absinthe on animals, and we have seen that this substance, when given in a weak dose, determined the occurrence of attacks of vertigo with muscular shocks, whilst in a stronger dose, it excited regular epileptic attacks. It gives rise, moreover, to a delirium with hallucinations, which, whilst it resembles that of alcohol in its general character, has a different mode of evolution; for whilst alcohol requires to prepare its ground before producing an outbreak of delirium, absinthe can immediately, at the first onset, produce in the animal, hallucinations with attacks of fright, anger, outcries, and agitation. This speedy appearance of delirium under the influence of absinthe, we find also in man, though not with the same abruptness as it affects animals, yet occurring sufficiently rapidly for the alcohol combined with the absinthe in the liqueur not to have had time to excite motor disturbances; this fact is well shewn, clinically, by the cases observed by M. Motet, especially his first case, mentioned at page 28. But the most important characters, and those which establish a radical difference between the action of absinthe and of alcohol, are furnished by the disturbances which take place with respect to the motor functions.

The muscular shocks which a weak dose of absinthe excites in animals, and even the vertigo, might pass unnoticed in a man saturated with absinthe and alcohol at the same time, the latter agent giving rise to a trembling which may mask the muscular shocks, and to attacks of giddiness which are somewhat analogous to those of vertigo; but, nevertheless, these symptoms will not escape observation on attentive examination. (Cases x., xi., xii.)

If the intoxication becomes complete, there appear unmistakeable symptoms, visible to everyone, I speak of the attack of epilepsy.

The attack of epilepsy is not, as many physicians think, the highest expression of the disturbance of motor functions in alcoholics, it is not the most extreme degree of that general tremulousness which is seen in *delirium tremens*, but it is a symptom of a different order which is superadded to the other motor phenomena. To be convinced of this, it is only necessary to note the circumstances under which the attacks are produced. Sometimes they sur-

prise the patients at the onset, sometimes at the
decline, of the alcoholic delirium, when the trembling is but slightly marked; at other times, on the
contrary, they shew themselves at the middle of the
attack, when the trembling is at its maximum; but
whatever be the time of appearance of the attack,
the trembling remains the same after it as it was
before. It subsequently diminishes if the attack of
alcoholic delirium is on the decline, whilst on the
contrary it increases, if the fit has come on at its
commencement. In Case ix., borrowed from the
thesis of M. Lagarosse, we read that on May 13th,
"The disturbance of movement has diminished,"
then on the 14th, "The night has not been so good
as the preceding, the patient has been excited; *an
epileptiform seizure.*" A little further on: "The improvement has not failed, the epileptiform attacks
have diminished, and since May 20th have not returned." It is not then a question of trembling,
but rather of attacks preceding the trembling, and
which, moreover, themselves disappear. There is,
then, no correlation between the attack and the degree of trembling.

In Case x. of M. Motet, on the 17th, we find
noted, "Trembling of the limbs;" on the 18th,
three fits occur, and as a motor disturbance we
find, "Extreme weakness of the lower limbs;"
there is no longer any trembling. On the following day a fresh attack occurs, and again only weakness of the limbs is noted without any mention of
trembling, which had no doubt almost disappeared,
since M. Motet, who was particularly watching the
motor troubles, does not speak of it. It is just the
same, moreover, in the other cases. When fits of
epilepsy come on in the course of violent attacks

of *delirium tremens*, one finds in the antecedents of the case the abuse of absinthe, so that one may say in a general way, "no fits, no absinthe," with the exception, however, of some rare cases in which the patients have abused white wine, bitters or vermouth.

According to the statistics of 1870 and 1871, which were communicated by myself and M. Bouchereau to the Medico-Psychological Society in 1870, of 377 alcoholics admitted to our asylums, 31 had had epileptic fits following the abuse of absinthe. In 1871, of 291 alcoholics, 15 had had epileptic attacks. These attacks of epilepsy, connected with absinthe, appeared both in patients with acute and with sub-acute symptoms, which abundantly proves, that, as we have already stated, there is no relation between the epileptic attack and the degree of trembling.

The peculiar action of absinthe is a fact the knowledge of which is not a matter of indifference, and perhaps bleeding would have been used more moderately in patients affected with epilepsy wrongly called alcoholic, if practitioners had been impressed with the idea that the attack was due to the direct action of a poison which possesses the peculiar property of stimulating, in a determinate manner, the excito-motor functions of the cord and medulla, without the necessary intervention of that *determination of blood* to the brain, which by some authors has been made to play the principal part in the production of convulsive seizures. Physiological experiments shew, moreover, that the cerebral hemispheres take no part in the production of the convulsive attack, which, under the influence of absinthe, occurs with the same characters in animals previously deprived of the lobes of the cerebrum.

It is needless for us to point out, that in the investigation of convulsive accidents in alcoholics, it is indispensable to separate the epileptiform convulsive seizures of chronic alcoholics (which are analogous to those found in general paralysis, senile dementia, and in patients affected with cerebral tumours, &c.), from the regular attacks of epilepsy observed in acute alcoholics with which we were just now occupied.* These latter depend on the special action of a poison, whilst the former occur as the result of profound lesions already produced in the nervous centres.

We may add that these epileptic attacks, which are very severe in some rare cases, usually terminate favourably. The grave cases are those in which the attacks are repeated, one after another, with intervals of only 5 or 10 minutes; at the third or fourth attack the person may succumb at once, or may remain in a comatose condition which ends in death; when, on the other hand, two or several hours separate the attacks, one usually sees them pass off without accident. The same gravity of the case is observed in animals when the attacks succeed each other rapidly.

The following cases, to which it would be easy to add a considerable number of others, will be sufficient, we think, to give clinical proof of the special action of absinthe and of absinthic epilepsy.

* Dr. Anstie, however, observed epileptiform seizures in some animals into which he injected alcohol. They occurred, however, only in rats and mice, and in these only just before death; they may therefore have a different signification. *Stimulants and Narcotics*, p. 401 (Trans.)

CASE IX.* M. A——, aged 33, of no profession, was admitted to Charenton on May 9th, 1863.

M. A—— has had a very good education. Since he left college, he has passed his life in idleness. Of a quick and hasty temper, he frequently abandoned himself to violent outbursts of passion. He passed his life in taverns and cafés, and indulged in copious libations. He was especially *a great drinker of absinthe.* For several years his vigorous constitution had resisted these excesses, but six months ago signs of alcoholic intoxication began to appear. The hands were agitated by trembling, especially in the morning when the stomach was empty; and at about the same time appeared convulsive attacks which still exist at the present time: These attacks, which, at their onset, occurred only every fortnight, have gradually become more frequent until they are now of almost daily occurrence.

The delirium with which he is now affected dates only from about eight days ago, and has only arrived by degrees at its present intensity under the influence of repeated libations. There are very active hallucinations of sight and hearing; he is pursued by his creditors, accused of theft, threatened with the gallows, &c.

There is very active delirium at night, hence the nightly terrors which led to his arrest by the police of whom he demanded help against his enemies. He has had but little sleep since the onset of the delirium, and there has been complete insomnia for the last three days.

May 9th. M. A—— is lying in bed in the infirmary, the disorder of the movements being such that he can hardly stand. Trembling of the hands; *subsultus tendinum;* face red, excited, eyes injected. The left eye presents a subconjunctival ecchymosis of some extent, but there is no trace of any blow which could have produced it; the skin is warm and covered with perspiration; the pulse full and frequent, 90 per minute. The delirium is not very violent, and it is easy to fix the attention of the patient; we can, however, discover the existence of hallucinations. The threats which he hears cause him much alarm.

10th. M. A—— has passed a very bad night with violent delirium, and very active hallucinations, he cried out and tried to escape from the dangers surrounding him. It was necessary to fasten him to the bed; he has had a convulsive attack, which, judging from the report of the attendants, closely resembled an attack of epilepsy.

* Case ix. is borrowed from J. Lagarosse, *Essai sur la folie alcoolique aiguë.* (Thèse de Paris, 1864, Case ii.)

The patient is less excited just now. There are the same symptoms as regards motility; pulse 95. Bleeding to 500 grammes (17½ ounces); nitre drinks; the blood drawn becomes covered with a thick buffy coat.

11th. The night has been bad, and equally restless as the preceding; bleeding to 500 grammes; nitre drinks. The patient was seized with a convulsive attack during the day at which we were present; *it simulated most perfectly an epileptic attack;* the blood is still covered with a buffy coat, but less than yesterday.

12th. The patient is calmer, face not so red, pulse 80.

13th. The night has been quieter, and the patient has had a little rest; to-day he is calmer and gives us some details of his disease. The subconjunctival ecchymosis has nearly disappeared, only a yellowish discoloration remaining; the disturbance of movement has diminished, the face is not so red, the pulse 80.

14th. The night has not been so good as the one before; the patient has been restless; an epileptiform attack. To-day he is calm, depressed and as if plunged in stupor; the hallucinations are very active.

In the afternoon, he made an attempt at suicide by throwing himself head foremost on to the paved floor. In the evening he renewed his attempt, and some bruises on the cheek and forehead have resulted.

15th. The patient is calmer, but continues sad.

From this day, improvement has steadily progressed, the epileptiform attacks have diminished and since May 20th have not reappeared. M. A——, for some time longer was worried by hallucinations especially in the evening, but these at last disappeared; the same is the case with regard to the motor disturbances; and at the end of the month of May, we may consider him as completely cured.

He remained however for some time in the private asylum, and the cure became complete.

CASE X.* M. D——, aged 33, a wine merchant, is at the head of an important house; an active and clever merchant, he makes his customers consume as much as he can, and himself sets the example; every day he drinks from eight to ten glasses of absinthe, and white wine and brandy in varying quantities, sometimes his potations go so far as drunkenness; such is the life he has led for about four years.

* Case x. is borrowed from M. Motet, *loc. cit.* Case v. p. 40.

On June 13th 1858, after two days of drunkenness, he was taken with delirium, accompanied by trembling of the hands and feet. He sees dirty little animals running about on the walls of his room and on his bed; he feels violent electric shocks; and believing himself the object of persecutions on the part of various persons, he tries to go out into the street, becomes excited and then in a passion; loquacity, incoherence and hallucinations are present.

He remains thus for three days, on the fourth the excitement abates a little, and M. D—— is taken with very copious perspirations; the trembling of the limbs is more marked; and there are sudden attacks of pallor.

We saw him for the first time on June 17th. There are extremely abundant perspirations, trembling of the limbs, a necessity for movement; these are real attacks, during which the patient follows an irresistible impulse, he throws himself about, and runs round and round; but there are no convulsions. Pulse 96, full, regular. In the night there is the same excited condition, abundant sweats, loquacity, hallucinations, sleeplessness, and great thirst. Is ordered red wine, largely diluted with water, a litre in the night; cold broth.

On the 18th at 7 a.m. there was a very well marked epileptiform attack, after which the tongue became dry, and the pulse rose to 112; extreme pallor, no perspiration; an aspect of general malaise; at 9 o'clock, there was a second attack; at noon, a third; both of the latter very violent, but not lasting long. After these there was extreme weakness of the lower limbs; it was necessary to hold up the patient, who tried to raise himself; general condition such as to excite uneasiness. In the presence of these symptoms, and not wishing moreover to have recourse to blood letting, it was resolved to place the patient in a bath, under the continuous fall of a stream of cold water; and the patient was left thus for four hours. In the evening at 6 o'clock, the heat of head had considerably diminished, the pulse was lowered, and the countenance less anxious; M. D—— listens when spoken to; he is put to bed; no sleep during the night, hallucinations of sight and hearing. On the morning of the 19th he has another fit. A bath for five hours, during which sensible improvement occurs; M. D—— answers questions; he does not yet know either what is the matter with him or where he is, but he knows that he has drunk a good deal. A basin of soup is given him whilst in the bath, which he drinks with pleasure. He is rather calmer on coming out of the bath; but he thinks a crowd of little animals are walking over him and go into his mouth; he is still very loquacious; he stands with difficulty, but still his whole general condition is more satisfactory. He is fed with soups.

20th. A better night, some sleep; the day calmer, but as yet no return towards sane ideas.

21st. Has a more exact appreciation of his condition. M. D—— feels that he is ill; there is less trembling of the hands, but still weakness of the legs. A bottle of Seidlitz water.

22nd. A large number of boils appear all over the body; the intelligence is becoming clearer.

25th. Loss of appetite; general malaise; tongue white; nausea and vomiting. A bottle of Seidlitz water.

27th. Condition very satisfactory; sleep calm, night undisturbed by hallucinations. Cure.

He went out on July 1st, 1858, perfectly recovered.

CASE XI.* *Excesses in wine and brandy; excess in absinthe; alcoholic delirium; epileptic attack.—Development of the same symptoms in three successive attacks—Albuminuria—Pneumonia.*

Louis Cl——, æt. 32 years, grocer and wine merchant, came into the Bicêtre on October 31st, 1863.

He is of good constitution, of nervoso-lymphatic temperament, of excellent health, and was of mild and uniform temper until the beginning of 1862, at which time changing his occupation, he ceased to be a pastrycook and became a grocer and wine merchant. Up to that time his conduct had been regular, and he had not committed excess of any kind; his hereditary antecedents present nothing noteworthy; he himself has had four children, two of whom have died from accident, but the two still living enjoy perfect health, the youngest is about 2 years old; their birth therefore was antecedent to the time of his setting up as a wine merchant.

From the month of February, 1862, Cl—— in order to improve his business and attract customers, made a point of drinking with them, and every day he took a rather large quantity of wine or of brandy, and from time to time a few glasses of absinthe. This sort of life had hardly lasted three months before Cl—— had already lost his appetite, slept badly, and became irritable. Then, to use his own expression, he " plunges into absinthe, to give him-

* Maguan, *loc. cit.* (*Union Méd.* August 4th, 1861).

G

self tone," and he confesses to having drunk from four to six glasses a day; but his wife states that he often took some secretly. This habit once contracted, the counsel of his friends and the entreaties of his wife do not prevent him from taking his glasses of absinthe; he feels better after having drunk them, and he feels something wanting, he says, when by any chance he has not made his accustomed potations. Under this diet, his health continues to deteriorate; the appetite, which was almost gone, becomes capricious; digestion is painful; retchings and sometimes vomiting of sour mucous fluid appear in the morning, and sudden faintings come on which oblige him to stop suddenly for a moment or two and run to sit down; there is also trembling of the hands which is most marked in the morning and rather less after meals.

These phenomena persist during the year 1862, they increase with each excess and diminish a little when the patient, over-tired and obliged to keep his bed, breaks through his deplorable habit for a time. But, at the beginning of 1863, Cl——, whose health is much impaired, "in order to recruit his strength," has recourse, even more frequently than before, to absinthe liqueur. Fresh symptoms soon appear; he is suddenly seized with a convulsive attack in church, during a funeral ceremony; he falls suddenly, loses consciousness and moves his arms and legs convulsively; his face becomes distorted, foam appears in the mouth and the tongue is bitten. After some minutes he comes to himself, has a stupefied look all day, but regains his usual aspect by the following day. From this time, the "absences" (or "faints") occur more frequently, but always without actual loss of consciousness. His health becomes worse, and about the 15th of October, as a sequel to fresh excesses in drink, in which absinthe still played the most important part, a second attack similar to the first comes on whilst he is going up stairs, and is accompanied by loss of consciousness, convulsions, foaming at the mouth, and biting of the tongue. The fall produces several contusions on different parts of the body and a wound on the orbital region. Some fatigue, and a stupid condition, continued for two days, then Cl—— began drinking again and was brought to the Bicetre on October 30th.

He remained excited, and cried out all night. In the morning, at our visit, we find him in bed, where he is kept by the straight jacket; he is pale and puffy looking; has a leaden tint, his face is covered with sweat, eyes brilliant, pupils dilated, equal; tongue white, deeply lacerated on both sides at the edge; abdomen hard, with constipation; urine of a reddish colour, when treated with nitric acid and heated gives a deposit of albumen of nearly half its bulk. Sensibility everywhere increased both to simple touch and pinching; he has considerable trembling of the legs and arms, and of the lips and tongue; especially when the latter is drawn out of the mouth; the voice is weak and the speech hesitating and tremulous. He is restless, raises his head, looks around him, changes his conversation every moment, is incoherent, has no idea of where he is, thinks he is in the street, at home, in his shop; thinks he sees his wife and children, speaks to them, tries to make them run away and escape from danger; then turning round sees rats and a spider at the foot of the bed; sees flames, is frightened and cries. He passes the day in the midst of these torments. He is ordered a large bath lasting for two hours, and a draught of 30 drops of laudanum.

Nov. 1st. No rest during the past night. The hallucinations persist, in the morning he sees the face of his daughter covered with filth. He complains of pain in both sides of the chest, respiration is frequent; no cough; percussion and auscultation reveal nothing abnormal. Skin warm, bathed in perspiration; pulse frequent, intermittent; the beats of the heart are irregular and succeed each other by jerks. Urine not so red; when treated with nitric acid and heat it gives a deposit of albumen not exceeding a third of its bulk. No pain in the region of the kidneys.

Prescription; a draught with thirty drops of laudanum.

Nov. 2nd. Has slept a little during the night, notwithstanding the persistence of the hallucinations; is calmer this morning, but retains a frightened aspect; answers questions slowly, and recollects in a confused way what he has been able to see or hear; knows where he is; there remain considerable trembling of the hands, and general muscular weakness, and the walk is insecure. There is less heat of skin, which continues moist; the pulse is more regular; respiration easier; pains in the side diminished; the urine gives with re-agents a deposit of about one-fifth of its bulk.

Prescription. A bath; draught of 30 drops of laudanum.

Nov. 4th. He slept during the night; countenance more restful; answers clearer; he inquires about his family, and speaks of his business, which he is anxious soon to resume.

The urine has become clearer, but still gives some flocculi of albumen with nitric acid and heat. The draught with the 30 drops of laudanum is continued.

Nov. 9th. The health improves every day; he has more appe-

tite ; the bites on the tongue are cicatrized, which enables him to
eat more bread and meat. The tint of the skin continues pale,
but the face is less puffy ; trembling of the hands persists. The
urine still gives some traces of albumen.

Nov. 11th. Nights good, sleep without dreams; answers clear
and sensible ; trembling of hands less ; little trembling of speech ;
countenance not so pale ; aspect composed ; appetite better, digestion regular. Urine no longer gives any deposit of albumen.

Nov. 15th. Improvement continues. His friends have been to
see him, and have told him of the state of his business, which is
going on badly; he is uneasy about it, thinks about it a good
deal, and wants to go out soon to look after it. The urine gives
a slight precipitate with heat and nitric acid.

Nov. 20th. Marked improvement; the trembling of the hands
grows less every day; the voice is clear ; speech not tremulous.
The countenance is resuming its natural expression, and is gaining
a little colour. The digestive functions are regular ; sleep calmer ;
urine does not give the slightest precipitate.

Cl—— stays in till the 23rd of November, at which time he is
allowed to go out on account of his business, although his health
was not as yet altogether re-established.

Second attack. Cl—— is brought back to Bicêtre on April
28th, 1864. According to the account given, he has never ceased
drinking since he went out; he took less brandy, but addicted
himself constantly to absinthe. He has got drunk several times,
and at those times become irritable and struck his wife. His constantly tottering health has made him incapable of all work ; and
he has kept his bed for three weeks, for an attack of inflammation of the lungs. It is now five days ago, that after having
indulged again in absinthe he had a convulsive attack resembling
the two attacks which he had last year.

On his entry, he is very pale, puffy-looking, his lips violet, the
sclerotica and the labio-nasal furrow yellowish, eyes suffused, and look
restless; he hunts about everywhere, goes round his bed, lifts up the
clothes, clings to surrounding objects, takes hold of his neighbours
and holds them forcibly, then lets them go, then goes towards them ;
goes about in every direction and seems to be the sport of numerous hallucinations of sight and hearing ; answers questions incom-
pletely, and cannot fix his attention for an instant; he pronounces
two or three words, then turning round looks alarmed, and mutters some incomprehensible words. His walk is uncertain, on the
least touch he gives a start and turns sharply round quite scared.

He has heard during the night his wife and children; some one has made them disappear into the stove, or into a basin; he does not know where they are. Sensibility exaggerated, and trembling of the hands and legs, lips and tongue; voice tremulous and slightly hoarse. Thirst; tongue red at the tip; epigastrium painful; no constipation.

Prescription. A bath; draught with 10 centigrammes (1½ grains) of gummy extract of opium.

April 30th. He has slept several hours during the night, has again heard the voices of his wife and children, he is downcast, oppressed, breathes painfully, coughs and brings up with great effort some mucous sputa which is dirty red in places, and slightly adherent to the cup; he feels a pain in his side, which is of a dull character, deep seated, and more felt on deep inspiration. There is comparative dulness on the right side in the lower two-thirds, more marked towards the base; respiration harsh behind, with fine subcrepitant râles; on the left side sibilant râles; skin warm; pulse 90, small, compressible. Urine slightly cloudy, and deposits, with heat and nitric acid, an albuminous precipitate occupying about one-fifth of the tube.

Prescription. Julep with 5 centig. of kermes; three pills containing 60 centig. (9 grains) of extract of cinchona.

May 1st. Countenance restless; sleeps badly; constant dreams; cough in fits, tiring; expectoration difficult; sputa mucous mixed with blood; dulness at the right base, with bronchial breathing; fine crepitant râle heard on deep inspiration; skin dry and warm; pulse small, compressible.

Prescription. Julep, with 20 centig. of kermes; three pills containing 80 centig. of extract of cinchona; a blister.

May 2nd. Aspect more restful; drowsiness; cough frequent with easier expectoration; sputa viscid, transparent in parts, more uniformly rusty; dulness on the right side, especially at the base; bronchial breathing; vocal resonance and fremitus more marked; subcrepitant râles through the rest of the right side; little warmth of skin; profuse sweats all over the body; pulse 94, fuller, but still compressible; urine clear; when treated with heat and nitric acid, becomes a little cloudy.

Prescription. Draught with 30 centig. of kermes; three pills containing 80 centig. of extract of cinchona; infusion of camomile.

May 3rd. Fatigue; sleepiness; less oppression; little pain; cough with easy expectoration; mucous sputa, yellowish, slightly adherent to the cup; some dulness on the right side with bronchial breathing; skin moist; abundant perspiration all night; answers calmer; has a confused recollection of the circumstances which led to his admission into the asylum; trembling of the hands; speech less embarrassed.

Prescription.—A draught with 30 centig. of kermes; three pills containing 60 centig. of extract of quinquina.

May 4th. Sleep during the night; aspect more restful; lucid answers; cough less frequent, with expectoration of mucous sputa, thicker, yellowish, some white and frothy; dulness diminished on the right side; respiration less harsh, subcrepitant râles; on the left, sibilant and mucous râles; nothing abnormal in the urine. Prescription repeated.

May 5th. Depression; malaise, sleep restless, with constant dreams; the chest symptoms are improving, but there is no appetite; mouth clammy, tongue white; abdomen slightly distended, diarrhœa with six stools a day.

Prescription. Draught with 2 grammes of diascordium: barley water with syrup of quinia.

May 8th. Weakness, indifference, drowsiness; no more hallucinations or dreams, pains in side gone, cough slight, with easy expectoration of yellowish mucous sputa; appetite poor; diarrhœa persisting.

Prescription. Two half injections with 20 drops of laudanum, and 1 gramme of extract of rhatany.

May 12th. Marked improvement; pallor and puffiness of the face diminished, sleep, appetite; less diarrhœa. The patient gets up for part of the day. The same prescription.

May 20th. Cl—— is in a good way of recovery, his sleep is calm, the appetite is returning; the face gains colour; his conduct is more natural; the cough has ceased and so has the diarrhœa. The hands and tongue no longer tremble; his answers are precise and intelligent.

He remained under treatment for about ten days longer, and then went out, having decided to become a pastry-cook again; and promising strictly to abstain from drink.

Sequel of the case. Third admission. Having gone out for the second time, Cl—— gave up absinthe for some time, but after a few days he began again to drink wine and brandy. His sleep became bad, hallucinations of the usual painful character re-appeared; he lost his appetite; a quantity of phlegm was brought up in the morning, and the limbs became tremulous. This condition lasted two months, when Cl—— finding himself weaker, had again recourse to his private drinking.

The absinthe soon excited fresh attacks of epilepsy. Cl—— entered the Bicetre again for the third time, on Dec. 5th, 1864. At the time of his admission, he still bore the marks of the bites on his tongue, in striking proof of the last fit.

Such is this case, not to say this experiment. Its subject is a vigorous man, who up to that time had been entirely free from every symptom of alcoholism, and all convulsive phenomena. He begins with excesses in wine and brandy, and becomes alcoholic; then he addicts himself to absinthe and becomes epileptic.

After his first stay in the asylum, the symptoms disappear. Then as soon as he has gone out, he begins to drink again, alcoholic symptoms occur first; he takes absinthe and there is a fresh attack of epilepsy. He stays in the asylum, and the symptoms cease. For the third time, there is indulgence in wine and brandy, and alcoholism occurs, then excess in absinthe, and epileptic attacks are superadded. A stay in the asylum and sobriety, and the symptoms cease.

CASE XII. *Excessive drinking, especially of absinthe—Alcoholic delirium, epileptic attack—Comparative temperance, diminution of the symptoms—Fresh excesses—Alcoholic delirium of maniacal form, speedily followed by delirium of melancholic character—Epileptic attacks.*

Louis D——, aged 40, formerly a soldier, but now a carpenter, was born of healthy parents, and his brother enjoys good health. Having become a soldier he was sent to Africa, where he contracted the habit of drinking. He was, he says, waiter at the canteen, and gave himself in the morning the first ration of absinthe. He had nightmares; he saw frightful things at night, and sometimes it happened to him, he said, *to make water in his bed, and to bite his tongue, and find his bolster spotted with blood in the morning.* Later, during the day, he had some epileptic attacks, and went into the hospital for an attack of delirium tremens. On leaving it he continued to drink, but in a more moderate manner.

His health had improved, and he had had no return of the convulsive attacks, when, three years ago, his brother says, he began to drink again both wine and brandy in considerable quantities. Fresh symptoms soon appeared, he used to get up in the night sometimes, looked about in his chamber, opened the door or window, and complained of imaginary enemies. He no longer had any appetite, he felt a loathing for meat, he ate little and drank the more as he felt his strength failing.

He had two attacks of epilepsy, occurring after an interval of several days.

Of late, he was restless during the day and worked very irregularly; at last, eight days ago, delirium came on with excitement and hallucinations which caused him to be put under restraint. He came into Sainte-Anne on April 20th, 1872, with the following medical certificate:—Acute alcoholic delirium, violent excitement, fears, cries and insomnia. Illusions of sight, terrors. Trembling of the limbs.

On his arrival D—— is in a state of extreme agitation, incessantly in motion, striking against the walls and doors, hunting on every side, bends down to the floor, looks at the ceiling; thinks he is in prison; imagines he has committed crimes; sees assassins and armed men pursuing him, &c. Profuse perspiration, trembling of the hands.

The night passes without sleep, he turns his bed upside down; looks and searches in all the corners; cries out, quarrels; answers insults; appears aggressive at times, at other times, on the contrary, beseeching and full of alarm. With the return of day, his attitude changes, and we hardly recognize the patient of the night before.

He holds himself erect, motionless, leaning against the wall; his aspect one of consternation, his eyes cast down: he does not answer when questioned, allows himself to be moved about, or offering only a very slight resistance, stopping when he is no longer guided and pushed. The most active efforts to get his attention do not succeed in distracting him from his pre-occupations, which show themselves from time to time by the expression of fear in his look and countenance. Whilst questioning him, he places his hands on the upper part of his thighs, then stoops down in a half bent posture and keeps this position for more than a quarter of an hour, grasping his leg firmly the while. He does not eat, if one puts food into his mouth, he does not move but rests motionless, with his mouth half open ; he swallows liquids, such as broth and milk, when carried by a feeding cup with a long spout to the back of the mouth.

Pulse regular, 70 ; skin slightly moist. Trembling of the hands much less than last night.

Treatment. A bath ; a cup of lemonade with wine ; a pot of broth and one of milk.

During the daytime there is no alteration in the melancholic and half stupid attitude of the patient; he is seated on the bed-chair, where he stays more than half an hour before making water, at a second attempt he passes a soft stool. One can get no answer, but his delirium appears less active and more change-able; on two or three occasions during the day, he carries his hand towards the lower part of his stomach and stays doubled up on himself. During the night he does not stir, but utters sighs at times, and does not sleep.

On the morning of the 22nd there is the same melancholic atti-tude ; and at last he says: "Don't kill me, I've done nothing wrong ; don't cut it off." Tongue dry, pulse regular, no heat of skin ; slight trembling of hands.

Treatment. Two bottles of vinous lemonade ; beef tea and milk.

During the day, we succeed in getting him to take more nour-ishment ; his countenance is less concentrated, but he does not speak. Night better, he sleeps several hours.

On the 23rd, at the morning visit ; he is sad, does not answer at first, but by degrees gains resolution to speak. He thinks he has insulted and outraged some women ; and he is condemned to a frightful punishment by the judges; he is to be eaten up by wild beasts, and he sees snakes ; he is to undergo mutilation : he felt yesterday, he says, the knife cutting him ; and heard his com-rades and mocking voices applauding the act.

In the daytime, he begins to recollect himself, and talks with those around ; he still hears some threats at night at the moment of going to sleep, he sees luminous points and white balls passing before his eyes.

24th. He is more tranquil and does not seem to have any hallucinations now, he narrates at length the strange fears he has undergone, and begins to doubt the reality of all those terrifying scenes.

The improvement continued on the following days as regards

the delirium, but in the course of May, two epileptic attacks occurred, and another on the 3rd of June; but without fresh intellectual symptoms.

The case of D—— is instructive, both from the
point of view of the special action of absinthe, and
also as showing the secondary importance to be attached to the form of the delirium, whether maniacal, melancholic, or stupid, which alcoholic delirium
may assume.

D—— whose hereditary antecedents present no
peculiarity as regards epilepsy, is in good health,
until, having become attendant at the canteen, he
addicts himself to brandy and absinthe; then there
quickly come on the intellectual disturbances peculiar to alcoholism in general, and, moreover, convulsive symptoms which we cannot but attribute
to absinthe. D——, in fact, was not an epileptic,
he was in the beginning of an attack of alcoholism
when the first attacks occurred, and their frequency
followed with tolerable regularity the more frequent
excesses in absinthe.

As to the maniacal, melancholic, and stupid forms
of alcoholic delirium, we find them associated
in D——, bearing only a relation to the greater or
less intensity of the hallucinatory disturbances, and
not leading to the least change in the nature, cause,
prognosis or treatment of the disease.

CASE XIII. *Excess in brandy and absinthe—vertigo and attacks
of epilepsy—attack of alcoholic delirium—persistence of some somatic
disturbances.*

Baptiste B——, aged 42, a coachman, admitted to St. Anne on
June 6th, 1872. There is nothing of importance in his hereditary
antecedents. For a long time past he has been given to excess in
drink, he drinks especially brandy and absinthe. Has been married four years, has no children; his wife had a miscarriage six
weeks ago.

For some years past he has been subject, at first to attacks of vertigo and muscular twitchings and then to fits. All of a sudden, a cloud passes before his eyes, and he is obliged to lean against something; for a second, he says, he is not there; on several occasions he has almost fallen from the seat of the carriage.

In his fits, he falls suddenly, loses consciousness, bites his tongue, throws his arms and legs about, and sometimes passes his urine and fæces; is at first pale and then becomes bluish; his lips are covered with foam.

For nine months past he slept badly, was frightened and had nightmares; he spat, and vomited greenish phlegm in the morning.

For the last few days, the insomnia was complete, he remained wide awake, he heard various persons; he thought he had murdered a man; he was accused of all sorts of crimes, he tried to hang himself.

On his entrance to the asylum, he is agitated, seems terrified, thinks he is an assassin; thinks he sees his wife drowned; sees himself surrounded by horses which are kicking. and fighting. Skin warm, moist; pulse full, compressible; considerable trembling of the hands, with sudden spasms at times in isolated groups of muscles.

Treatment. A basin of milk; two cups of barley water.

June 12th. He has had an attack of vertigo during the day, he was seated on a chair, when he turned pale, his arms fell, he stayed motionless for an instant, then he joined again in the conversation, having only a very vague notion of what he had just passed through.

His nights are still disturbed by hallucinations and nightmares; in the daytime he is more tranquil, occupies himself usefully and helps the attendants in their work in the ward; but he still appears irritable and discontented, and bears his confinement with difficulty.

Heaviness of head and numbness of the limbs; slight trembling of the hands, little appetite.

Treatment. For three days, one of Anderson's pills is given with his evening meal.

The patient continues to improve, his sleep is no longer disturbed except by dreams; in the day he works, and is steady, but retains a certain degree of irritability. From time to time he has pains in the head and cramps in the calves. He promises to become more sober and to leave Paris, so as to avoid temptations; he leaves on July 10th.

This patient presents nothing in his hereditary antecedents which could be connected with an epileptic affection, he drinks absinthe early, and is first affected with muscular spasms and attacks of vertigo, and then with epileptic fits. Such cases are very numerous, and on examining authors, we find vertiginous and convulsive symptoms described, which they have ascribed in a general way to alcoholism, but which, from the accounts themselves, are evidently due to absinthe. In this patient, notwithstanding the somewhat rapid disappearance of the hallucinations and delirium, there remain pains in the head, numbness and cramps, which disclose a somewhat advanced degree of poisoning, which, however, is certainly susceptible of cure by the cessation of excessive drinking.

ARTICLE V.

Treatment.

I. *Drunkenness.* The first indication to fulfil in a drunken man is to promote vomiting if we have any reason for supposing that any spirituous liquid remains in the stomach.

We may use tartar emetic, or still better powdered ipecacuanha, the action of which will be enhanced by drinking lukewarm water; or again, tickling the soft palate, which is more simple, and can be used when rapid action is desirable, may be sufficient to excite vomiting. Afterwards, we recommend rest in the horizontal position, the head being slightly raised and the body sufficiently covered to avoid the action of external cold, which when added to the coldness due to the drunkenness is often the cause of complications. We may also use liquid ammonia, with advantage, administered either as a draught or as an enema. Tea, coffee, stimulating liniments, the application of sinapisms, &c., are also means which are found of service in cases of weakness, and of more or less marked general torpor.

II. *Alcoholic delirium.* Alcoholic delirium, which as we have seen comprises the sub-acute forms of alcoholism, or cases of simple delirium tremens without fever and without complications, requires few agents derived from the materia medica, but rather prophylactic measures capable of preventing the return to excesses.

As pharmaceutic means, we may resort to some purgatives, and more rarely to emeto-cathartics to fulfil indications relative to the digestive tract, and to relieve the *primæ viæ* of the catarrhal products which so frequently accompany excess in drinking.

In some cases enemata will be necessary to overcome the obstinate constipation which occurs in some alcoholics. Diuretic and diluent drinks of mildly aperient nature should be freely given for the first few days; and later on, these may be very advantageously replaced by weak decoctions or infusions of quassia amara, gentian, calumba, or hop, taken during the day and at meals, during which they may be taken mixed with some wine instead of wine and water.

These bitter drinks apparently aid digestion, and we have seen them substituted for brandy and too copious rations of wine by some topers to their great

advantage. Lastly, coffee or a little quinine wine
taken uniformly after meals, not before, and especially not in the morning on an empty stomach, may
be used with profit and will perhaps be accepted
later on by the patients as their ordinary and sufficient stimulant. Some drunkards sincerely believe
that they need a stimulant, that their disordered
stomach is incapable by its own powers of fulfilling
the work of digestion, and they have recourse
to strong drinks to give it tone and vigour; they
cannot be convinced by reasoning, but as they sometimes feel the benefit of these bitter drinks they
finish by forming the habit of taking them.

Our excuse for entering into details of apparently
so little importance is found in the necessity there
is for insisting on all means, little or great, which
combat drunkenness.

Confinement in an Asylum. But, besides the therapeutic treatment which finds its means in the materia
medica, there is another very important question
well worthy of engaging our attention, I speak
of sequestration or confinement, and the duration
of this confinement.

And, firstly, is it necessary to confine all
patients attacked with alcoholic delirium? Put in
so general a way, this question could never be
answered. It is just for the solution of these eminently practical questions that a classification of
persons affected with alcoholic delirium into distinct
groups is indispensable.

As to the second and third groups which we have
recognized :—(the second, patients with slow convalescence and ready relapses; the third, patients predisposed to delirium); it is evident that sequestration
is an absolute necessity ; but with regard to mem-

bers of the first class, in whom convalescence is mild, rapid and complete, and also with regard to a certain number of patients in whom the delirium disappears at the end of three or four days, one may fairly ask if it is indispensable to make them pass through an asylum.

M. Lasègue ends his article on sub-acute alcoholism by the following sentence:[*] "Lastly and always because it is only a transient symptom, sub-

---

[*] Lasègue *Arch. Gén.*, August, 1869, p. 163.

M. Lasègue's definition of sub-acute alcoholism. "By sub-acute alcoholism I mean the form which is apyretic, free from great perturbations of the central nervous system, coma, maniacal fury, &c., and from the local or general disturbances of the circulation which accompany acute alcoholic fever; of a duration rarely exceeding a fortnight and almost always less than a week; it is characterized by a delirious condition so peculiar, that by it alone we may be able to affirm the nature of the malady, by a trembling equally characteristic, by insomnia, and by more or less marked digestive disturbances; a form defined with sufficient exactness to warrant a prognosis, although it comes midway between real acuteness and chronicity; and so common as to answer to three fourths of the cases which come under our observation." (*Loc. cit.* p. 157.)

Under the designation of sub-acute alcoholics, M. Lasègue appears to us to join together facts which from a practical point of view should be separated. On the one hand, in fact, we are dealing with patients who may be well in less than a week, and on the other hand with patients who may not be cured at the end of 15 days; now sub-acute alcoholics whose delirium persists beyond fifteen days, often have notions of being persecuted, melancholy ideas, &c., which make them dangerous and which consequently distinguish them from other alcoholics. I ought to add that occasionally it is impossible, when the fit of alcoholic delirium comes on, to foresee if it will be followed by delirious ideas of a certain duration, and moreover, that at the end of the attack the latter are not always easily appreciated. We should therefore be liable to disappointment by not taking precautionary measures, or rather by giving these patients their liberty too early, in cases where their sequestration would have been effectual.

acute alcoholism ought always to be recognized by
the physician with sufficient certainty for him to
affirm its nature, and its early cure. To confound
it with madness which will last some time, and
which demands *decisive measures,* would be to commit
an awkward mistake whose dreadful consequences
I need not to enumerate."

M. Motet,* in his analysis of M. Lasègue's article,
formulates the practical conclusion to be drawn from
this work. It is difficult to estimate, says he, all
the importance of the practical deductions which
should be derived from it, the most weighty of all
is undoubtedly this: "It is not necessary in the immense majority of cases to send individuals attacked
with sub-acute alcoholism into asylums for the insane; their cure is rapid, and in order to obtain it
such isolation and repose as can be secured for them
otherwise than by a legally-authorized sequestration
is sufficient." The idea is certainly a good one,
much better treat an alcoholic whose delirium is of
less than a week's duration elsewhere than in an
asylum for the insane, but the disappearance of hallucinations and of delirium is not the complete restoration to health, more often, we know, these
patients retain for some days longer a condition of
cerebral malaise, their sleep is disturbed by dreams
and nightmare, and there are gastric disturbances
which urge them to fresh excess and to a speedy
relapse.

So that in reality, in the actual state of things,
as we do not possess refuges or special hospitals in
which to keep this class of alcoholics long enough
for their complete cure, it is preferable in our
opinion to send them to asylums, where, moreover,

* Motet. *Ann. Médico-psych.,* July 1871, p. 108.

thanks to constant oversight, surprises are less easy, and the patient is restored to liberty when he is really found to be in a condition to enjoy it without abusing it.

What we have said about the second group of patients, those affected with alcoholic delirium of mild convalescence and easily relapsing, will be sufficient to shew the necessity of a prolongation of their stay in the asylum so as to guarantee their cure for some time, even after the cessation of the delirium and hallucinations. We know in fact that these patients tend to chronic alcoholism, that with fresh relapses irremediable structural disorders will be produced in the organism; it is necessary, therefore, by a more protracted sequestration, to place them under conditions the most favourable to the maintenance of their health, and to the resistance of the external and internal incentives which prompt them to further excesses.

On the other hand, we frequently find in this class of alcoholics, patients who are led on to drinking by their business, such as cooks, wine merchants, certain tradesmen, workmen, who having left their country to increase their salary, are lured by example in the workshops, and contract drunken habits; for these patients, safety lies in giving up their excesses, and the most urgent exhortations and most weighty advice can do nothing to change their sad habit; it is absolutely necessary to remove them, and to take them away from the midst of that which tempts and incites them. We have sometimes succeeded in obtaining this result, three times we have learnt from patients some months after leaving Paris, that the means had answered; but more frequently we have received no news, and probably their ex-

H

cesses have been resumed; some indeed, by their readmission into the asylum have proved to us how frail was their resolution.

As to the third group, which comprises patients with a predisposition to delirium, in whom relapses are frequent, and whose attacks of alcoholic delirium are followed by delirious ideas which persist for a long time, we can only do good by prolonging their confinement, and by not allowing them to leave until at least three months or even more than that after the complete disappearance of all delirium. Besides, a review of the observations made on these unfortunate victims of alcohol will readily convince us, that, sad to say, the time of their residence in the asylum is the least painful part of their existence. But surely there are prophylactic measures which we should no longer neglect for avoiding fresh relapses after they have left the asylum, and in a still more general manner for preventing drunkenness. As far as concerns our three groups of patients suffering from alcoholic delirium, there is one of them, the first, and fortunately the most numerous class, which is susceptible of improvement and of cure, to which we must address ourselves, and towards which all our efforts must be directed. For to drag an alcoholic from his fatal habit, is not only, as we know, to restore an individual to health, but it is to put him in a condition to procreate other than imbeciles, epileptics, or fools.

The prophylactic treatment of alcoholism has been the constant object of the thoughts of hygienists, and M. Jolly, after having pointed out the various pharmaceutical means, adds; " But it must be thoroughly understood, that the moral treatment ought to take the precedence of all pharmaceutical and other treatment. As a pro-phylactic, it is needless to say that all means are summed up in sobriety; in sobriety which is not abstinence, but moderation wisely applied according to conditions of age, sex, constitution, climate, and which, it is necessary to say, is governed much more by example than by precept; consequently, example as a principle of physical and moral education, example during infancy, ex-ample in the family, example in all social circles, here is the true prophylaxis of alcoholism; and it is in this direction especially, that the Executive or even legislation can efficiently interfere in the name of hygiene and public morals."*

Under the inspiration of such thoughts as these, the Academy has taken a happy initiative in the crusade against drunkenness, and the "*popular treatise*" on the dangers following the abuse of alcoholic drinks which emanated from the able pen of M. Bergeron has been the starting point of the efforts which have been attempted in this direction.† Two temperance societies have been already formed, the one having at first had for its only object to war against the abuse of tobacco has enlarged its scope, and is become "the French Association against the abuse of tobacco and of al-coholic drinks;" the other is "the French Associa-tion against the abuse of alcoholic drinks."‡ Allow me to reproduce four paragraphs of the statutes

* P. Jolly. *Etudes hygiéniques et médicales sur l'alcool et ses composés. Bull. de l'acad. de med.*, 1865—66, vol. xxxi, p. 490.

† Bergeron. *Avis sur les dangers qu'entraine l'abus des boissons alcooliques. Bull. de l'acad. de med.*, July 25, 1871, vol. xxxi, p. 528.

‡ A. Foville. *Moyens pratique de combattre l'urogneric*, Paris, 1872.

of this latter society, in which are enumerated the principal measures by which it seeks to realize its object. It proposes;

a. To favour the replacement of alcoholic liquors by wholesome drinks, such as natural wines, cider, coffee, tea, and beer; especially by means of co-operative societies for the sale of refreshments.

b. To promote the formation of workmen's clubs, where the members shall find honest and useful recreations, and from which spirituous drinks shall be excluded.

c. To try to obtain an increase of the tax on alcoholic drinks, and as much reduction as possible on other drinks, having at the same time a due regard to the interests of commerce and industry.

d. To demand new preventive measures against drunkenness, particularly, a diminution in the number of taverns, and a strict regulation of all traffic in drinks.

But, in addition to these means, which are directed chiefly against immediate dangers, we must follow out the work of regeneration and moralization, without which all other efforts would give but insufficient results, and we say with M. Chauffard: "We must instruct, and above all, infuse good morals into this generation. Let us thoroughly understand that instruction alone will remain powerless. It is useless merely to teach the people the deadly effects of alcoholism, of which, moreover, they are not ignorant, they will none the less allow themselves to be carried away to the full measure of their gross enjoyments. The masses of humanity have never been restrained by a sense of their interests. To check men in the face of their passions, we must imbue them with ideas of morality, of duty and devotion, of self-abnegation and sacrifice.*"

* Chauffard. *Bullet. de l'Acad. de Méd.*, vol. xxxvi. Session of Jan. 25th, 1870.

CHAPTER III.

FEBRILE DELIRIUM TREMENS.

ARTICLE I.

Symptoms and Diagnosis of Febrile Delirium Tremens.

THE study which we have thus far made of the acute symptoms of alcoholism has shewn us this intoxication under a favourable aspect as regards the prognosis of the attack of alcoholic delirium, yet nevertheless, most of the cases hitherto related, might have been placed in the class of what is generally considered under the name of *delirium tremens.* But we are about to see that there are nevertheless sufficiently well marked differences between the latter and simple alcoholic delirium to allow of a very distinct separation between them. In both cases, we find the same toxic delirium. This may, it is true, present an activity and acuteness in *delirium tremens,* such as one seldom finds in simple alcoholic delirium, but frequently also, febrile *delirium tremens* is accompanied by delirium of moderate intensity; thus the patient, Auguste L—— (Case xiv.) gave us a sufficiently sustained attention to allow of the exploration of the general sensibility, and of the senses, and the sensibility of the retina; in his case also, we often received clear and precise answers just as in cases of simple alcoholic delirium. The patient Jules L—— (Case xvii.), likewise shewed only a moderately severe delirium, notwithstanding the presence of grave symptoms from the first day. We cannot then base the differential diagnosis upon the intensity

SYMPTOMS AND DIAGNOSIS. 103

of the delirium. And it will be useful to observe as regards this subject that, speaking generally, the activity of the delirium and the amount of cerebral excitement accompanying alcoholism is far from being in direct proportion to the dose of poison; the blow is struck alike in all cases, but the instrument does not always vibrate in the same manner. The special delirium is never wanting, it occurs in all alcoholized persons, with its hallucinations of a painful and changeable character, based either upon the daily occupations, or the subjects of predominant interest at the time, but what great varieties do we find in the mode of reaction, varieties of which the sole cause is the subject acted on!

This peculiar influence of the subject acquires, as we have seen, its greatest proportions in predisposed patients. Frequently, in this class of alcoholics, a violent delirium follows excesses in drinking, without the intensity of the symptoms bearing any relation to the quantity of liquor they have taken, and after the more acute symptoms have passed off, there still remain as we know, various delirious ideas. And it is a remarkable fact, that these persons, who are subject to very frequent relapses, are often brought back to the asylum with violent attacks of alcoholic delirium, accompanied by intellectual phenomena of an extremely acute character, which contrasts strangely with the mildness of the physical symptoms; they have slight tremor, a moist tongue, a moist skin, and a countenance which gives rise to no anxiety; the storm, in fact, usually passes over without any accidents. It is necessary to keep these facts before one's mind, in studying febrile *delirium tremens*, in order not to ascribe to the intellectual disturbances an importance which they do not possess, and especially not to conclude from their mildness that the prognosis is also favourable.

Since the delirium can give us only insufficient information, or may even lead us astray in some cases, let us seek in the physical signs those which may serve as a basis for diagnosis and prognosis. Some have insisted on the injection of the eyes, the alteration of the features, the profuse sweating, and the frequency and force of the pulse, as characters of grave delirium tremens, but as these symptoms may be met with in simple alcoholic delirium, they cannot be sufficient for the purpose of diagnosis. The really characteristic signs are the following; first and most important, that which will give the real signal of alarm, is the indication drawn from the element of *fever*, which we can now appreciate in an exact manner, and which we may in some sort calculate with the thermometer in our hands. In every case where, in a patient affected with acute alcoholism, and free from any thoracic or abdominal complication, in a word, without any intercurrent affection which is in itself capable of exciting and keeping up fever, every time, we say, that the temperature taken in the rectum, after having oscillated during two or three days about 39° C (102·2° F), rises to 40° (104° F.), or 41° (105·8° F.), the prognosis will be grave, and its gravity will augment with the increase or even the persistence of this elevation of temperature (Cases xiv, xv, xvi.) When, on the other hand, after a somewhat rapid elevation to 39°, or even to 40° as in the patients A——— and C———, (Cases xix and xx) one sees the temperature go down at the end of 24 or 48 hours, the prognosis ceases to be unfavourable, and soon all the symptoms taken together enable us to foresee a favourable result. In the case of acute alcoholism of mild form, the mercury may rise to 38·4° (101·1° F.), or 38·6° 101·5° F.), in consequence of the agitation, at times extreme, of the alcoholic, but this limit is rarely exceeded, and in all cases, as soon as the patient has a moment's respite, the column rapidly descends, and oscillates about 38° (100·4° F.)

The pulse, as is seen from the tracings, does not follow the course of the temperature in a regular manner, and cannot furnish us with any precise information. The indication afforded by the fever, is not the only one which serves to guide us, we can draw a second which is also of great value from the disorders of movement, and here too we must distinguish the intensity itself, and the activity of these motor disturbances, as well as their duration, persistence, and degree of generalization. An alcoholic, for instance, may present an enormous trembling of the arms, legs and face, but if this trembling does not persist, or if it does not affect all the muscles of the body, we may reassure ourselves, it is only a stroke, a passing symptom which will often abate, without leaving any injurious traces of its occurrence; but if the trembling, which is otherwise of moderate intensity, has affected all the muscles of the body, if it is accompanied by shakings, and especially quiverings and muscular undulations, and if all these phenomena persist, without disappearing during sleep, not only do we arrive by about the second or third day at nervous exhaustion, and complete prostration of strength, but the persistence of all these symptoms is the sign of an intense irritative process going on in the nervous centres, and especially the spinal cord,

an irritation of which the autopsy unfortunately
too often gives proof; these two signs, which are
of such real importance, and so easy to recognize,
may nevertheless, unless care be taken, pass unnoticed.

The skin, in fact, which is often moist with perspiration, only allows of an incomplete appreciation
of the temperature by the touch, and hence it is absolutely necessary to have recourse to the thermometer. On the other hand, if the muscular quiverings,
the undulations, and slight shocks, are but slightly
marked in the face, we often do not think of applying the hand to the body, and thus feeling under
our finger, the cries of suffering of the spinal cord,
whose excito-motor forces exhaust themselves in vain
in this useless, tumultuous and disordered action.
The patient, Auguste L——, even at the moment
of answering our questions, shook thus from head
to foot; and the hand laid on the thighs, or the
muscles of the back, felt, on making slight pressure, undulations and quiverings, together with
shocks even in the deeper parts. If the attention
is confined to the simple inspection of the surface
of the skin, we see, no doubt, the little elevations
and depressions, and small waves produced by the
unequal and irregular contractions of the bundles
of muscular fibres, but in reality we do not appreciate the intensity of an action so continuous and
so general.

Lastly, we often see muscular weakness supervene. The patients, who for the first two or three
days can stand upright and walk in spite of the
general tremulousness, then begin to totter, and
their grasp is feebler, although the trembling may
not have increased. This incomplete paralysis,

which is usually most marked in the lower limbs,
is often sufficiently explained by the hyperæmia
which is sometimes so considerable, and by the
more or less extensive hæmorrhages which are
disclosed by the autopsy, not to mention the
special action of the poison, which has by this
time acted particularly on the spinal cord, and has
rendered the influence of other causes more active.
This third sign, based on the state of the muscular
powers, is of less importance than the two previous
ones drawn from the temperature and the muscular
quiverings and shaking, because it is not constant,
and especially because it is of later appearance.

We will not refer again to the part played by
meningeal hæmorrhages in the production of pachymeningitis, (Cases xvi, xvii, xviii), we have seen that
by themselves, they may sometimes be sufficient to
produce false membranes.   Nor need we insist upon
the lesions found in most of the organs, which are
usually due to fatty degeneration, with the exception of the condition of the liver of the patient in
case xviii, in which, according to M. Calmeil's de-
scription, cirrhosis was found. This author says that
the parenchyma of the liver was granular, and of
yellowish colour.   On section, with a bistoury, it
was resistant and resembled a condensed fibrous
tissue.

It remains for me to speak of the presence of
alcohol in the substance of the tissues, which we
have already had occasion to notice, in the physiological part of this work at page 11.   In case xiv,
the analysis of the organs, made with care, yielded a
negative result.   Auguste L—— had left off drinking seven days before his death.   In case xvii, the
liver and brain contained alcohol, but none was

found in the blood or in the other organs. The patient, Jules L——, at the moment of death, had passed three days and six hours from the time of his last excess in drink. In another patient who died fifty hours after his last potations, alcohol was found in the liver, brain and blood in very appreciable quantity; the lungs and kidneys also contained traces. And, lastly, in another autopsy where a chemical analysis could not be made, but the liver emitted a distinct smell of alcohol, the patient had died forty hours after his latest excess.

We know that MM. Lallemand, Perrin and Duroy have found alcohol in the organs of a man who had succumbed to the usual complications of drunkenness, thirty-two hours after having drunk a great quantity of brandy, and that at the time, the direct effects of intoxication had disappeared. At the present time several observers since MM. Lallemand, Perrin and Duroy have determined the presence of alcohol in the organs of animals killed during drunkenness.

According to these different results, the presence of alcohol naturally in the system during the days immediately following the last excesses is incontestable. In one case, as we have seen, there was still alcohol in the liver and brain at the end of three days and six hours; in another case at the end of seven days, there were no traces of it left. Physiological experiment would be necessary to fix the limit between these two times. However it may be, we know that the poison is remaining in the tissues and organs, at the most acute stage of the poisoning; and hence we may draw an indication of great importance to act upon, one which must not be forgotten, especially in the case of patients with febrile *delirium tremens* who come into asylums, who most frequently are admitted after numerous recent excesses.

We have indicated at the end of each of our cases, the special points which seemed to us most worthy of remark, and we shall not return to them here; we must, however, point out the existence in the patient C—— in case xix, of a paralysis of motion and sensation, which was incomplete and generalized, but most marked throughout the whole of the right side. We shall have occasion to return to this interesting fact in the article on chronic alcoholism, where we shall find other cases of hemiplegia with hemi-anæsthesia of general sensibility and of the senses.

CASE XIV. *Father a drunkard—Excess in wine, brandy and absinthe. Two epileptic attacks. Febrile delirium tremens. T. in rectum* 38·8° *on admission, rising to* 43° *at death.—Autopsy : Fatty degeneration of all the organs, more particularly the muscular system. —Chemical analysis discovers no alcohol in the viscera ; (the excess had ceased seven days before death).*

Auguste L——, aged 36 years, a pork-butcher, whose father was a drunkard, enjoyed good health during his childhood; when 18 years old he began to drink, and soon attained the habit of taking wine, brandy, and liquors of all kinds.

He lost appetite and sleep, and was compelled several times to keep in bed for some days. For the last four or five years he has vomited phlegm every morning, and has frequently had pains and cramps in the stomach ; to cure these he has increased the dose of white wine and absinthe. For three years he has suffered from headache, giddiness, pains in the arms and legs, together with a weakness of the lower limbs, which has at times obliged him to sit down. For a year past he has had trembling, and has not done his work so well; he is sad and listless, complains of being restless at night and of not sleeping, in conversation he appears distracted at times, as if he were thinking about something else, and sometimes he seems to be looking for somebody.

For some months now he has been irritable, often in a passion,

and drinking furiously.   Lately he has eaten scarcely anything, trying to obtain food seasoned with vinegar and highly
spiced.   In 1870 he appears to have been seized with a convulsive
attack whilst he was on guard on the ramparts; about this time
he occasionally had most violent shakings in the arms and legs.

On the 29th of June he appeared to be restless and pre-occupied
in thought, he said that people bore him ill-will, that thieves
were seeking to rob him; and in the day-time he was seen to
be taking his trunk and other things into the cellar in order to
place them in safety.   On the 30th he suddenly lost consciousness
whilst in his shop; he fell to the ground, writhed, bit his tongue,
foamed, and passed water in his trousers.   In the evening his
fears were ceaseless, he saw people on the roofs of the houses;
muskets pointed at him, he was surrounded by a swarm of flies
which buzzed about his head, got into his eyes, his ears, his nose,
and his mouth; his spittle is full of balls composed of insects.
He had a bad night, the restlessness continued the next day, and
he came into the asylum on July 3rd.   On admission his hallucinations were incessant, he was speaking, crying and constantly
on the move.   The night passed without sleep.

July 4th.   At the visit, the hallucinations persist with the same
intensity; L——— picks up objects on his clothes, walks carefully
so as not to upset vases which he sees on the ground, shakes dus-
ters which are covered with filth; answers his customers, speaks
to his master, hides himself and wishes to run away to avoid the
bullets.   Trembling of the whole body, quivering of the muscles
of the face; eyes injected.   Skin warm, tongue furred and deeply
bitten on the right side, pulse quick, compressible, 76. T. R.
38·8°C. (101·8 Fah.)

Treatment.   Two glasses of sedlitz water; two bottles of citric
lemonade; a litre (1¾ pints) of broth; two dishes of soup.   During the day L——— is calmer and one can easily fix his attention.

The general sensibility is not altered, taste appears to be dulled,
nevertheless, sugar and salt are recognized; the sense of smell
appears to be normal.   The retinal sensibility is altered, there is
dyschromatopsy.   The colour red is called chocolate; yellow is
called red and carmine; green is red; blue is white.

As soon as one ceases to ask questions, the hallucinations reappear, he draws thread from between his fingers, tears out thorns
which he sees bristling on his skin; thinks he is in the shop, sees
a head of pork, it must be salted, he says, or it will be spoiled; he
sees artillery-men marching on the wall, and Chasseurs de Vincen-
nes, quite an army.

The trembling persists in the arms and legs, with shuddering and quivering in the muscles.

July 5th. Sleepless and restless all night. In the morning the hallucinations are of the same intensity and excite loquacity, cries, and fear. Quivering of muscles all over the body; profuse perspiration; pulse 80, small and difficult to count because of the subsultus tendinum. T. R. 40·6°C. (105° F.)

Treatment. Wine mixed with sugar and water; milk; broth; three dishes of soup. A bath.

At 5 o'clock in the afternoon, is talking incessantly; his sentences broken and referring to his business. L—— talks about legs, sausages, then he wants to drink, "I pay you for a pint"; then wishes to take his musket and exercise—he must join his battalion, "attention, attention." Sentences short and abrupt.

L—— begins to stagger, cannot hold himself upright; up to this time he had been able to be out of bed, and to walk quite easily; when seated on the bed his body trembles and cannot remain still a moment; all the muscles of the body in addition to the trembling and transient shakings are seized with quivering and undulations easily appreciated on applying the hand.

Tongue moist, lips dry, nostrils clogged, eyes bleared and injected; perspiration profuse. The pulse, counted with difficulty at the radial, gives 108 pulsations at the femoral, it is quick, compressible, and at times irregular; the heart's beats are a little indistinct; there is vesicular breathing on each side of the chest with some scattered râles; resonance normal. T. R. 40·6°C. 105°F. (the same as in the morning at 10 o'clock).

Midnight. We find the patient lying on a mattrass, covered by his clothes, and sleeping profoundly; his breathing easy and regular; his face bathed in sweat; his face is coloured but not cyanotic; the trembling continues in the arms and legs, and the feet, which project a little beyond the mattrass, present a rhythmic and quite regular trembling. His sleep is not disturbed during our visit.

July 6th. Hallucinations less frequent and referring more particularly to his daily occupations. The patient answers some questions, he knows he is in the hospital, but has only a confused recollection of what has taken place; he appears to be very low-spirited.

The muscular quiverings and tremors do not cease, they are less intense but always general. Pulse 89, T. R. 40·8°C. (105·5° F.)

Treatment. Milk; broth; wine with sugar and water; extract of cinchona, one drachm; syrup of orange, seven drachms; three dishes of soup.

The patient having been a little quieter since morning, answering some questions well enough, and shewing less of the trembling, was seized at 3 o'clock with slight convulsive twitching in the arms and legs, attended with rigidity of the neck and contortions of the face; deviation of both eyes to the right, and a little foaming at the lips. This convulsive attack lasted scarcely a minute and was repeated half an hour subsequently. A semicomatose condition succeeded the second fit.

At six o'clock respiration was embarrassed, and abdominal; the face and lips blue, the eyes dull; the small and wretched pulse beat 120 in the femoral; the hands and feet were cyanotic and cold. Temp. raised to 42·4°C. (108·3° F.)

The trembling continued, but with less force; by scratching the soles of the feet some reflex action could be obtained.

8.30 p.m. Coma, resolution of the whole body; cessation of the trembling, respiration suspirious, and followed at the end of expiration by a rather loud laryngeal sound; cyanosis and coldness of the extremities, of the nose, ears, hands and feet; corneæ glazed. Pulse thready. T. R. 43°C. (109·4° F.)

Died at 10 p.m.

Autopsy at 6 a.m., on July 8th.

The urine having never been examined during life, the patient having passed his water several times under him, about 60 grammes were drawn from the bladder, it was thick and yellowish, and on being warmed became milky; treated by nitric acid it left a deposit of albumen about one-fifth. Bareswill's solution produced no change.

Brain. The cerebral dura mater is slightly injected, the arachnoid and pia mater are œdematous especially towards the summit of the hemispheres, and at several points they adhere to the dura mater at the site of the Pacchionian bodies. Over the middle part of the hemispheres on both sides, the infiltration is slightly rose-coloured; at some points are seen little ecchymotic spots, in size never exceeding that of a lentil. The membranes are detached quite easily.

The cortical layer, slightly rose-coloured throughout its whole extent, presents a red dotting at the convexity of the hemispheres in the parts corresponding to the infiltration of blood in the pia mater. The grey portions of the optic thalamus and corpora striata are likewise rosy. The vessels at the base of the brain are not atheromatous. The fourth and lateral ventricles are smooth and free from granulations. Sections made in all directions reveal no special lesion in any place beyond the injection of the grey substance.

Fig. 1.—Auguste L——, æt. 36.

Febrile Delirium Tremens.

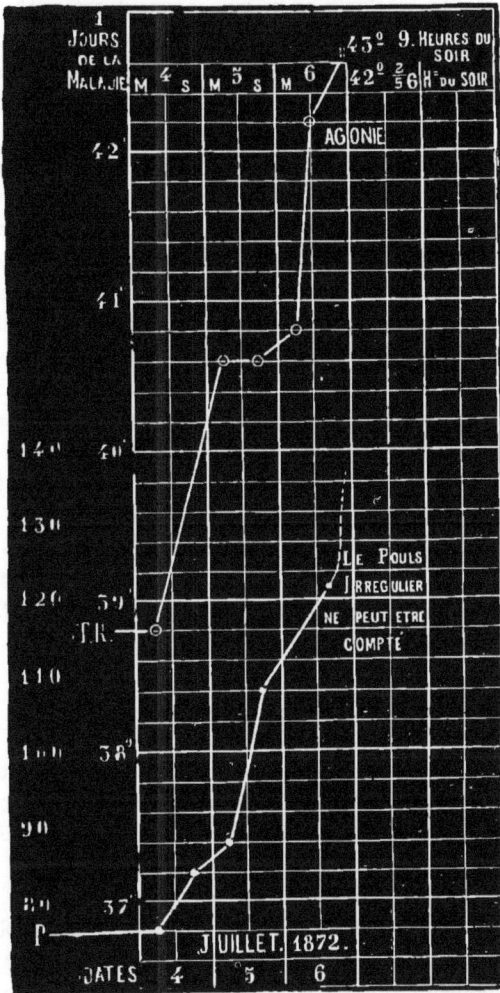

EXPLANATION OF FIG. 1.—The first column represents the number of beats of the pulse per minute, indicated by P and the line passing from it. The second column indicates degrees centigrade of the temperature taken in the rectum. The remaining columns indicate morning and evening temperature on the 4th, 5th, and 6th of July, which correspond with the 4th, 5th, and 6th days of the attack. The temperature will be seen to rise to 42·4°C (108·3° F.); at 6 p.m. on the 6th; reaching 43° C (109 4° F.) at the time of death. The pulse from 6 p.m. is irregular and uncountable, and is represented as becoming more rapid till the time of death.

I

Spinal cord. The spinal dura mater presents a rosy tint, which is very marked in the upper portion on starting from the basilar groove, and from the periphery of the foramen magnum : this rose-coloured staining which affects the whole surface and does not disappear on washing, is in marked contrast with the paler colour of the cerebral dura mater. Right through the spinal canal, after removing the cord and its envelopes, we find a dark red tint on all the inner surface ; the sinuses are gorged with very black blood.

On slitting up the dura mater we see little venules coursing over the surface of the pia mater beneath the arachnoid, which are tortuous, injected, and ready to burst. Sections of the cord made in different regions shew everywhere a very marked and uniform injection of the grey substance ; in no place do we observe any extravasation, nor relative excess of the punctiform injection.

Liver. The organ is of normal size, its under surface is very green. On tearing off the capsule we see a smooth uniform surface, yellowish in some places, especially in the left half, where the staining, as is shewn by sections, penetrates more deeply. The gall bladder, which is moderately full, contains brownish-yellow bile.

The *spleen* is of moderate size and normal appearance.

The *right kidney* presents a greenish tint superficially, the surface of both kidneys is glossy and smooth ; on section the cortical portion is yellowish as well as its prolongations between the pyramids.

The *pericardium* is adherent to the heart, in tearing it off we break some small very loose fibrous threads.

The *heart* is soft and flabby ; the wall of the right ventricle is covered in front with a thin layer of yellowish fat ; the heart-tissue has a yellowish brown tint which is more marked in some points ; the walls of the left ventricle also present a yellowish colour, but less marked than on the right side.

The inner surface of the aorta is spotted in some places with small irregular patches of a yellowish-white colour, occupying the superficial layer of this tunic, as is ascertained by making sections through the walls.

The lungs, which are of a deep red colour, more marked at the base, crepitate, and pieces thrown into water do not sink.

The mucous membrane of the stomach is slightly injected ; the intestines shew no arborizations nor injection.

The mucous membrane of the bladder is healthy.

The blood from the venæ cavæ, superior and inferior, when collected in a vessel, shews very small, yellowish, shining drops, resembling oil-drops, floating on the surface.

The muscles in all parts of the body have a yellowish look; but this discoloration attains a considerable degree in the erector spinæ muscles; all the sacro-lumbar mass is of a very marked pale yellow. The muscles of the thighs, of the calves, and of the arms, (deltoid and biceps) are a little redder than the other muscles.

Microscopic Examination. Fine sections and preparations obtained by teasing the cortical portion of the brain, shew small shining granulations in places on the wall of the capillaries, some of which are isolated, others being agglomerated into small elongated masses. At some points crystals of hæmatoidin are found. The nerve-cells are infiltrated, and some are distended with granules of yellowish pigment, but in all the nucleus is easily distinguished. Preparations of the cord shew similar lesions; and on the walls of the capillaries of the grey substance we see a greater number of blood crystals.

Preparations of the liver derived from the most yellow portions shew a rather advanced stage of fatty degeneration, the cells are infiltrated and distended with fatty granules and oil-drops; we also see little drops and granules of fat lying free in the preparation.

In fine sections taken from the cortical portion of the kidneys the tubes are seen to be opaque, enclosing epithelial cells infiltrated with fatty granules, and also free granules outside the cells.

The muscles are the seat of a most marked fatty degeneration, but all the fibres are not affected to the same degree. In several preparations (and on this point we repeated our examination several times) we see lying beside a healthy fibre, marked with very distinct striations, and without granules either upon or inside the sarcolemma, a fibre in which the striation is not apparent, and which is filled with small granules and in some places with more voluminous granulations; at some points larger fat-globules are seen among the fibres. When treated with ether, the small granules become paler and partially disappear, and larger granules appear.

The muscular fibres of the heart present granules of yellowish pigment on the sarcolemma, some lying between the fibrillæ and having a linear arrangement, others infiltrating the

nuclei and grouped in little masses; the contents of the primitive
sheath, *i.e.*, the sarcous element itself, is less granular than in the
muscles of the trunk and limbs. This fatty degeneration of the
heart appears to us to be very like the fatty degeneration peculiar
to senile dementia.

CHEMICAL ANALYSIS OF THE ORGANS IN SEARCH OF ALCOHOL.

This analysis was made in the laboratory of M. Paty, chief
chemist to the St. Anne Asylum, by M. Beufvre, a distinguished
interne of the Asylum.

The viscera cut up separately, have about a fifth of their weight
of water added to them, and are pounded in a mortar so as to
form a thin pulp. Each of them with the blood contained in it
is placed in a tubular retort, which communicates by means of a
jointed tube with a receiver resting in cold water. The retorts are
heated by a sand-bath so as to produce a slow distillation. The
operation is arrested when a quantity of liquid of about one-
eighth of the weight of the material employed in each distillation,
has been collected.

This first step in the operation being completed we have recourse to benzoic chloride, which has recently been pointed out by
M. Berthelot as a suitable reagent for detecting the presence of
alcohol in liquids which contain only minute quantities of it.
The liquor obtained from distillation of the liver was first acted
upon, about thirty cub. cent. of the liquid being placed in a test-
tube and a few drops of benzoic chloride added; the tube being
then gently warmed and shaken. Under these conditions the
benzoic chloride seizes the least traces of alcohol and forms benzoic ether, which dissolves in the excess of benzoic chloride and
settles at the bottom of the tube. The supernatant liquid is de-
canted and on the remaining chloride a solution of potash is
poured, and the tube heated. The potash attacks the benzoic
chloride immediately and its odour disappears; but if benzoic
ether be formed, it reveals itself even when in a small quantity,
by a strong and permanent odour resembling that of essence of bitter almonds. In the liquid submitted to experiment, all ethereal
odour had disappeared after the addition of potash, which proved
the absence of benzoic ether, and consequently the absence of
alcohol. The other liquids procured by the distillation of the
brain, lungs, blood and kidneys gave equally negative results. On
the other hand, a comparative trial made on ordinary water con-

taining a few drops of alcohol gave a very sensible odour of benzoic ether. To verify these results by another test, we tried the action of these various liquids on chromic acid, using a mixture of bichromate of potash and sulphuric acid in the proportion of 1 to 30 by weight. On pouring into each of these liquids a few drops of this solution, shaking and warming it, the green tint which is easily obtained in water very slightly alcoholized, could not be got. We may, therefore, conclude that alcohol was absent from the viscera submitted to experiment.

Like several of our patients, L—— had taken brandy and absinthe to excess, and he presented alcoholic delirium attended with trembling and with attacks of epilepsy; further than this we have noted in his case sudden transient shocks which existed a year before the fit of delirium tremens and at the time at which the first epileptic attack occurred. The hallucinations and delirium of L—— were extremely well marked and turned upon his daily occupation; on several occasions he thought he was in his shop, he saw hams, a head of pork, &c., he wished to take his gun, to do exercise, this in remembrance of his service in the National Guard. This delirium, although rather active, presented, however, no excess until the twelfth day after admission: it was possible for us to keep his attention sufficiently alive to test the general sensibility, the senses, and the sensibility of the retina, and thus to ascertain his difficulty in judging of colours. Although the delirium had nothing special about it, yet as regards the muscular system we noticed a tremulousness which, if not very intense, extended to the whole body, and was moreover of such persistence that neither by day nor by night, neither in the state of quietness nor in sleep, did we ever see it cease for a single moment. In the last moments of death alone did complete stillness replace the trembling. But the important symptom and the one which from the second day inspired grave fear, notwithstanding the apparent improvement, was the progressive rise of temperature, which from 38°8 rose to 40°6, the next day to 40°8 and at last to an excessive height; from the moment in which the breathing became embarrassed to the death struggle which followed close upon the two slight epileptiform attacks.

At the autopsy we found nothing in the brain to explain the conjugate deviation of the eyes to the right, a deviation which manifested itself at the time of the epileptic attacks and which remained during the semi-comatose state. The fatty degeneration of the muscles had reached a degree considerably at variance with the degree of. steatosis in the other organs. Finally, chemical analysis gave a negative result, which is of so much the more importance as it marks one of the limits (seven days) at which alcohol ceases to be present in the human viscera.

In the case which follows, the chemical analysis will show the very manifest presence of alcohol in several organs of a patient who succumbed three days after the termination of all excess in drinking.

CASE XV. *Febrile Delirium Tremens—Accidental hæmorrhage; aggravation of the symptoms—Diminution of the trembling in the arms and legs; persistence and increase of the muscular quivering and of the shaking— T. R. 39° C. (102·2° F.), on admission, rising to 41°2 C. (106° F.), at death. Autopsy: considerable injection of the spinal cord; fatty degeneration of the viscera.*

Leopold C——, æt. 53 years, a waiter at a wine shop, came into the Asylum on March 15th, 1873, on the third day of an attack of delirium tremens.

16th. Slight excitement, hallucinations; he hears his relatives, he sees animals around him, and people who threaten him. One has to call loudly to get an answer from him. Trembling of the whole body; some shaking of the limbs. In the middle of the night there are cries and fears, he breaks the panes of a window in trying to defend himself against imaginary enemies, and causes several wounds on the arms and legs which bleed freely.

17th. Violently agitated, loquacious, incessant painful hallucinations; his attention cannot be fixed on any point. Trembling of the arms and legs, shaking and quivering of all the limbs; intense thirst, white tongue. Pulse 90. T.R. 39° C. (102·2° F.)

Treatment. Two glasses of Seidlitz water; vinous lemonade; broth; milk; two soups.

Evening. Delirium the same. Pulse 94. T. R. 39°6 C. (103·3° F.)

18th. Restlessness continues, hallucinations, he is busy with customers, his master, and his comrades; he drives off animals, especially insects. Trembling of the arms and legs; at times more severe shakings; diarrhœa; thirst less intense.

Pulse 116. T.R. 39°8 C. (103·6° F.)

Treatment. Vinous lemonade, broth, milk, chocolate; julep with 4 grammes (one drachm) of watery extract of bark.

Evening. Trembling of the limbs, quivering of the muscles of the trunk and thighs; pulse small, tremulous, 120. T.R. 40°4 C. (104·7° F.)

Nothing particular on auscultation.

19th. Less restless, he answers questions readily enough, but stops to address imaginary beings. The trembling is less, but the muscular quivering persists. Pulse 120. T.R. 40°4. The same treatment.

Evening. Pulse 130. T.R. 40°6 C. (105° F.)

. 20th. The patient is quiet, he no longer tries to get out of bed. He speaks to his friends and to his relations, he sees animals. One succeeds in gaining his attention, and in obtaining some sane answers. The trembling in the muscular masses of the arms and legs is but slight, but the quivering continues. Pulse 112. T.R. 39°6 C. (103·3' F.) The same treatment.

Evening. The patient is exhausted, retains some hallucinations and mutters incoherent sentences.

The trembling has almost entirely disappeared, but we observe subsultus tendinum, convulsive quiverings in the face, fibrillar movements and little shocks in the muscles of the trunk, the gluteal region and the thighs.

Fig. 2.—Leopold C——, æt. 53.   Case XV.

Febrile Delirium Tremens.

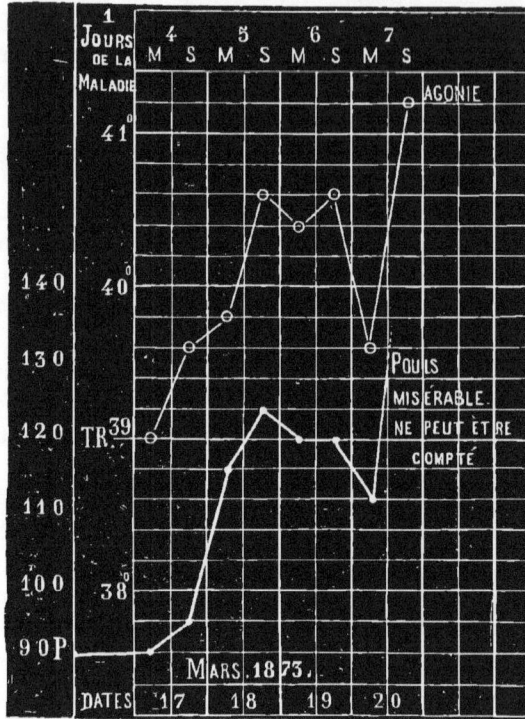

EXPLANATION OF FIG 2.—The first column indicates the number of beats of the pulse per minute, indicated by P. and the thick line passing from it. The second column, the degree centigrade of the temperature taken in the rectum. The headings of the remaining columns indicate the morning and evening of the days of the disease, corresponding with the 17th, 18th, 19th and 20th of March, respectively.

The rectal temperature, indicated by the thin line T.R., will be seen to have risen from 39° C. (102·2° F.), to 40·6° C. (105° F.), and after falling to 39·6° C. (103·3° F.) on the morning of the 7th day, it rose to 41·2°C. (106·2° F.) at the time of the death-struggle. The pulse after reaching 130 towards the evening of the 7th day, rapidly rises and becomes wretched and uncountable.

Erythema and commencing bed-sore on each buttock.   Pulse small, frequent, wretched, cannot be counted. The temperature taken in the rectum gives, 41°2 C. (106° F.)

The patient died in the night.

Autopsy. The cerebral meninges are infiltrated with serosity and are easily removed over the whole extent of the brain. The cortical portion is pale or slightly yellowish. Sections made in succession from the periphery to the centre of the hemispheres shew no hæmorrhagic lesions in any place. The lateral ventricles allow a little serosity to escape, but the ependyma is normal; the fourth ventricle presents nothing particular.

The spinal meninges are injected, especially towards the end of the dorsal region ; sections made at different heights of the cord reveal a considerable injection of the grey substance, which is the more marked, as we approach more nearly to the lower extremity.

Heart soft, loaded with fat on the surface, in the right half; pericardium milky at certain places ; walls yellowish on the right side, and of a yellowish brown on the left.

Lungs healthy with the exception of slight injection at both bases, but there are no hæmorrhages; the injected portions crepitate and float.

Kidneys yellowish throughout the cortical part.

Liver uniformly yellowish throughout.

Stomach, walls thickened, mucous membrane of a violet colour, turgid and ulcerated in the great curvature for an extent of about 10 centimetres (4 inches) ; at the circumference of the ulcer, are infiltration of blood and slight ecchymoses.

The heart, liver, and kidneys, when examined microscopically, shew an advanced stage of fatty degeneration.

It is worthy of note in this case how the accidental loss of blood which occurred on the second day aggravated the physical and intellectual symptoms; and, without ascribing too great an importance to this fact it nevertheless gives us a certain lesson with regard to the use of bleeding in delirium tremens, a mode of treatment, however, which is almost abandoned in our day.

It is useful also to notice the very decided independence which is shewn between the trembling of the arms and limbs found in ordinary cases of alcoholism, and the muscular quiverings and shocks with subsultus tendinum, which are the habitual accompaniments of serious cases; the trembling indeed in our patient had almost entirely disappeared by the third day, whilst the muscular quivering persisted and became more intense towards the last moments of life.

Lastly, the considerable injection of the cord contrasted so strongly with the almost normal condition of the brain that one cannot help recalling the predominance of motor over intellectual disturbances.

CASE XVI.   Pierre T——, æt. 40 years, a coal merchant, having been given to habits of drinking for a long time, and his faculties already being weakened, was seized with delirium tremens on Feb. 26th.   Taken to the Hotel Dieu, he was from thence sent to St. Anne on February 28th, 1873.

March.   Violently agitated, subject to hallucinations : he sees men, women, and soldiers who wish to shoot him; he sees swords and arms of all kinds.   People threaten him, wish to poison him, and to kill him.   There is trembling of the limbs.   Pulse 56. T.R. 39°2 C. (102·6° F.)

Evening.   Condition the same.   Pulse 56.   T.R. 38·8° C. (101·8° F.)

March 2nd.   Restlessness, loquacity, hallucinations.   Pulse 60. T.R. 39·2° C. (102·6° F.), in the evening.

March 3rd.   Hallucinations, loquacity, remains up, walks about in all directions.   Pulse 52.   T.R. 38·6° C. (101·5° F.)

March 4th.   Hallucinations, cries, ideas of poisoning, hypochondriasis, slight trembling, shaking of the limbs.   Pulse 56.   T.R. 40° C. (104° F.)

Evening; the pulse cannot be counted because of the movements of the patient.   T.R. 37·8° C. (100° F.)

March 5th.   Excitement, loquacity, trembling less, but muscular shocks and quiverings.   Pulse 80.   T.R. 40·1° C. (104·2° F.) Evening ; pulse 84.   T.R. 40·4° C. (104·7° F.)

March 6th.   Prostration, coldness of the extremities, commencing cyanosis ; hallucinations, incoherent talk.   Fibrillar movements and quivering of the muscles.

Pulse 76.   T.R. 40·4° C. (104·7° F.)   4 o'clock p.m.; pulse 84.   T.R. 40·4.

March 7th.   1 a.m.   Cyanosis, asphyxia; respiration 60, pulse 108.   T.R. 42·4° C.   Died at 4 a m.

At the autopsy, thickening of and infiltration of blood in the meninges were found ; punctation of the cortical portion is seen after removing the pia mater which adheres slightly.   Fatty degeneration of the heart, liver, and kidneys.   Thickening of the coats of the stomach ; marked injection of the mucous membrane.

FIG. 3.—PIERRE T——, ÆT. 40. CASE XVI.

Febrile Delirium Tremens.

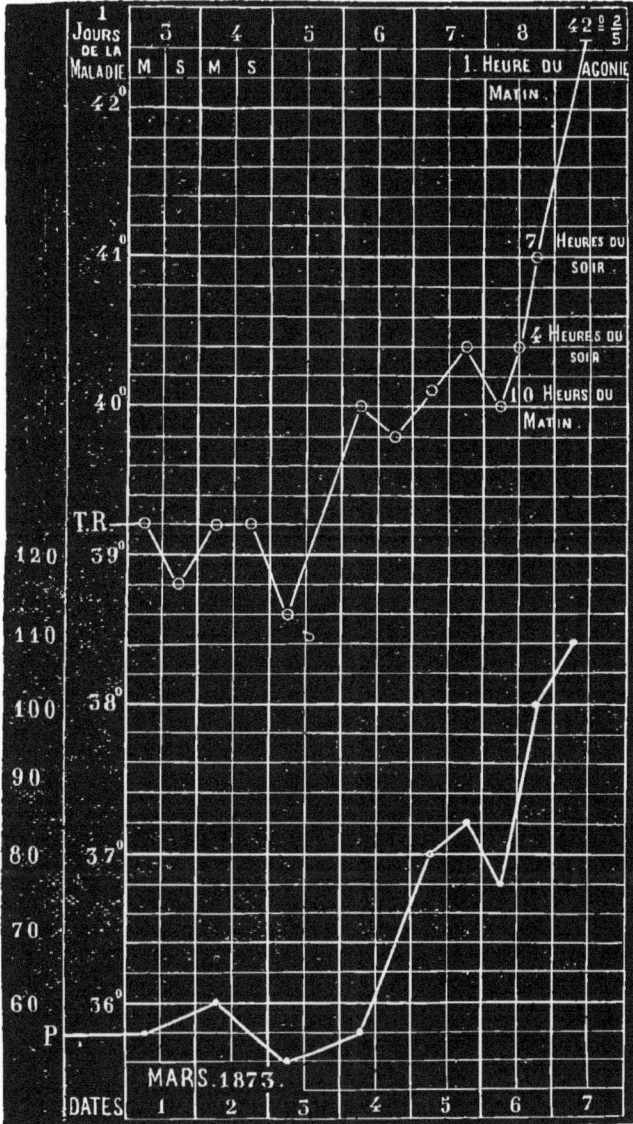

For explanation of the terms in the figure, see Figs. 1 and 2, pp. 113 and 120.

CASE XVII. *Alcoholic habits—White wine—Absinthe—Epileptic attacks.—Febrile delirium tremens—Trembling, quiverings and muscular shocks.—Autopsy—Injection of and hæmorrhage into the cerebrospinal meninges ; fatty degeneration of the viscera—Chemical analysis shews the presence of alcohol in the liver and brain—The excess ceased three days and six hours before death.*

Julius L——, æt. 24 years, a waiter in a wine shop, had been accustomed to drinking, and had committed more numerous excesses the last three years, taking white wine and sometimes absinthe. On the days immediately preceding his admission into the asylum he had been in several debauches, and had been dismissed from his situation, and on the eve of the day on which he was placed in confinement he had fallen on the staircase in an epileptic fit.

On admission on Sept. 19th, he was very excited, loquacious, and constantly on the move ; he saw insects, armed men who wished to strike him, he heard his relatives, &c. Countenance changed, eyes injected, tongue moist, skin warm and covered with sweat, pulse large, compressible, frequent ; trembling of the arms and legs, with quiverings of the muscles of the face.

Treatment : milk ; broth ; two bottles of citric lemonade.

Passes a very bad night, the hallucinations do not give him a moment's rest.

Sept. 20th. The hallucinations and delirium continue. L—— is constantly looking for and driving away insects ; he rolls up the bed clothes into a bundle and wants to be off ; he talks to his friends ; on speaking sharply to him we get some answers, but his attention is soon distracted by hallucinations. Profuse perspirations, muscular quiverings and shocks at times ; trembling of the arms and limbs ; a little muscular weakness, gait tottering.

Pulse 80, large, regular, T. in the rectum 39° C. (102·2° F.) No lung affection. Urine albuminous. In the forenoon he had an epileptic attack, followed by stupor of about a quarter of an hour's duration.

Treatment. Milk, broth, wine and water ; two bottles of citric lemonade ; three dishes of soup.

21st. The hallucinations continue, he is very weak and can scarcely stand ; trembling of the limbs, more marked in the arms ; quiverings and some shocks at times ; skin warm and bathed in perspiration ; pulse regular, frequent, and yields to the finger. Urine still albuminous. Treatment the same.

About 2 o'clock in the afternoon, L—— is exhausted and so prostrate that he can no longer sit up in bed ; he mutters incoherent sentences, and talks about his customers, his friends, bottles and corks. The trembling and quivering persist throughout the body, but are more feeble. The pulse is frequent, small, compressible. The skin is slightly moist.

Towards midnight, about a quarter of an hour after having given him something to drink, the attendant on approaching his bed found him dead, without having heard any movement or groan.

Autopsy, 30 hours after death. Marked injection of the meninges; hæmorrhage for the extent of about 2½ inches spread as a thin layer over the visceral arachnoid on the surface of the right hemisphere; on the left hemisphere is a little effusion disposed in the same manner, but of less extent. At the convexity of the hemispheres, the arachnoid and pia mater are infiltrated and of a rosy tint; at the base these membranes are almost as thin and transparent as in the normal condition. They are easily torn off except at the anterior part of the two frontal lobes, where they slightly adhere to the cortical portion, which at this point is of a rosy lilac colour : on the left sphenoidal lobe they are adherent over a small area. The whole surface of the cortical portion presents a lilac tint, which is more pronounced towards the anterior parts. The grey portions of the central part are also of a deeper colour than natural.

Sections shew no hæmorrhage at any place.

The pons and medulla oblongata are injected.

The spinal meninges, which are of a deep red, are infiltrated, and at certain places, either superficial to, or beneath the visceral layer of the arachnoid, in the thickness of the pia mater, they present little blackish clots of blood.

Sections of the cord shew a very marked injection of the grey substance.

Kidneys large : cortical portion yellowish.

Liver, a little enlarged, of a yellowish tint which is more marked in certain parts.

Spleen, normal.

Lungs of dark colour, congested ; they crepitate and float.

Heart, loaded with fat, soft ; the walls are of a yellowish-brown colour in some places.

The mucous membrane of the stomach is infiltrated with blood in two places on the great curvature of the size of a sixpence.

Chemical analysis, performed as in the preceding case, shewed the presence of alcohol in the liver and brain, and traces could be recognized in the lung. None could be found in the blood, although a very careful analysis of it was made.

Julius L——'s excesses were chiefly in white
wine, and sometimes he took absinthe; he had
two attacks of epilepsy, one on the day before he
was placed in confinement, the other on the day
after his admission into the asylum; the latter was
followed by stupor lasting a quarter of an hour.
The delirium was marked by the ordinary features,
but the trembling in this case affected the whole
body and was accompanied by frequent quiverings
and muscular shocks which denoted severe irritations of the cord.

General weakness soon appeared without any disappearance of the other motor disturbances, and the
patient died very suddenly without any pain.

The *post-mortem* shewed us multiple and disseminated hæmorrhages in the cerebro-spinal meninges,
and as far as the cord is concerned occupying all its
regions; and we can understand the seriousness of
disorders so wide-spread in such important organs as
the brain and cord. These severe lesions explain
the so frequent existence of chronic meningeal irritation in the more advanced stages of alcoholism; it .
is probable that if L—— had survived this attack
of delirium tremens, a kind of thorn, so to speak,
would have been left, which even apart from any
fresh excesses would have been quite capable of
provoking such serious chronic lesions, for instance,
as pachymeningitis, and all the symptoms with respect to motion and sensibility which result from
it.

Chemical analysis, by shewing the presence of
alcohol in the liver and brain three days and six
hours after the cessation of all excess in drinking,
teaches us the necessity of endeavouring before all
things, in these cases of delirium tremens, conse-

quent on recent drinking, (which is most commonly the case, at least as regards patients admitted into asylums,) of endeavouring, we say, to promote the elimination of the poison by all possible means, as its action will be just the more energetic as its stay in the system is the more prolonged.

We give the following case of M. Calmeil's, which is interesting on account of the marked and general injection of the cerebro-spinal membranes.

CASE XVIII. *Accustomed to use of fermented drinks—Reverses in business, an attack of apoplectic nature after a full meal, furious delirium for four days and four nights, death at the commencement of the fifth night—The bones of the cranium injected internally; the pia mater beginning to be infiltrated with serosity, and red in patches; the grey matter of the cerebrum and cerebellum rose-coloured, numerous large and empty vessels in the medullary substance, vertebral sinuses turgescent; the bones and dura mater and membranes of the cord are stained with blood around the prolongation of the canal.**

M. Audry, aged 49 years, born in the Seine et Oise department, married, a retail dealer in wine at Paris, has formed the habit for a short time past of eating little and drinking much; he has not however abused stimulating liquors to such an extent as to make himself really drunk. He is a man of great corpulence and of very firm muscle and has had to be bled copiously and very frequently in the course of each year. It appears that he readily gets diarrhœa; but the ailments he suffers from have seldom hindered him from attending to his business at the counter. His intellectual faculties are of ordinary ability, his manner is lively and abrupt.

In September 1840, he was subjected to great vexation by the decision of a valuation-jury that his house should be pulled down for the sake of public utility; he was greatly dissatisfied moreover at only having obtained an indemnity quite out of proportion to the loss they had compelled him to suffer; this cerebral excitement being kept up by the more and more frequent ingestion of fermented liquors into the digestive organs. On September

* Calmeil, *Traité des Maladies Inflammatoires du Cerveau (Treatise on the Inflammatory Diseases of the Brain)*, 1859, vol. I., p. 224; case 36.

13th he was at a grand breakfast, at which the guests drank with but little moderation, but from which M. Audry returned without any appearance of indisposition or illness.

However, on the 14th before day-break M. Audry was suddenly seized with an attack of cerebral congestion. The doctor who was immediately called in bled him freely as soon as possible, and the apoplectic phenomena soon disappeared; but scarcely was he restored to consciousness before an outbreak of symptoms of the most violent delirium occurred. The patient was now besieged with hallucinations of sight and hearing. He thought he both saw and heard workmen engaged in pulling down his house, and got in a furious passion with persons who tried to quiet him. He proceeded even to acts of the greatest violence against one of his relatives who was trying to hold his head and to hinder him from throwing himself out of the window: they then took the precaution of fastening him to his bed.

All the day through on Sept. 14th the loquacity, vociferations, abusive language, threatening, petulancy, and efforts to break the bonds which fix him continue. Fifteen leeches are applied, but not without some difficulty, behind each ear; acidulated drinks are given him.

On September 15th. the excitement keeps up at the same rate; ice-cold applications are continuously applied to his head; the lower limbs are enclosed in revulsives, and an endeavour is made to allay his thirst by the use of nitre-drinks.

September 17th. The features of his countenance are profoundly altered; all efforts made to execute the doctor's orders are more and more powerless, and as the impossibility of contending against such a malady is felt, M. Audry is sent to Charenton, where he expires at the end of three hours. His vociferations are continued even to death and he has not ceased for a single moment to struggle under the bed-clothes in his efforts to break his bonds.

Autopsy. The abdominal and thoracic cavities are well developed; the arms, thighs and legs are highly muscular. At the bend of the arm and in the neck are traces of recent cicatrices which shew that the bleedings have not been sparing. The top of the head is the seat of a large infiltration of blood.

The bones of the cranium present nothing particular with regard to thickness; they are injected on that part of their surface which corresponds with the dura mater.

The right cavity of the cerebral arachnoid having been opened first, we notice over the right lobe a condition of very considerable serous infiltration into the meshes of the pia mater. On the same side the visceral serous layer is raised by layers of blood which resemble wheals or extravasations of various size.

Over the left cerebral lobe the tissue of the pia mater is likewise the seat of a serous infiltration, but the posterior lobule is enveloped as it were in a reddish cloud, vascular threads very close together and highly hyperæmic appearing to constitute for the most part these islets of blood, as they may be termed.

The pia mater is everywhere easily detached from the convolutions.

On the right as well as on the left side it seems to be thickened; the surface of the brain is generally smooth, firm, and free from erosion, except in some places, especially on the right side, where the cortical-surface presents very small solutions of continuity which are, however, quite superficial.

The consistence of the brain is in no wise unusual.

There is a scarcely perceptible rosy tint of the inner layer of the grey-substance. The vessels of the white substance are very numerous, very apparent, but only moderately full of blood; nevertheless in certain places the junction of the vascular tubes produces a sort of reddish marbling of more vivid colour. The cerebellum is not soft, its grey substance presents in many points shadings of the colour of the dregs of wine.

The pons is firm, its grey substance resembles that of the cerebellum as regards colour.

The spinal plexuses of veins, and the veins of the cord are swollen and laden with black blood. This liquid has discoloured the face of the vertebræ and the ligaments.

The spinal dura mater is likewise of a red colour on its external surface throughout its whole extent, owing to the action of the colouring matter of the blood. The spinal arachnoid and the membrana propria of the spinal cord are, with respect to colour, in the same condition as the fibrous membrane and the tissue of the vertebræ.

The spinal cord is not altered; its structure and colour are quite natural.

Both costal pleuræ are of a rosy tint; the right pleura is adherent by some old fibrous bands.

Both lungs appear to be healthy externally, the bronchial mucous membrane is very red even to its termination in the pulmonary vesicles.

K

The heart, although large, is not considered to be diseased, bearing in mind the marked development of all the other muscles.

The right auricle contains large clots of fibrine; and its endocardium is of a violet colour. The interior of the aorta is no longer grey; on the contrary it is of a purplish colour, and its violet tint cannot be removed by washing.

The parenchyma of the liver is granular and of yellowish aspect: on cutting it with the scalpel it is resistant and resembles condensed fibrous tissue.

The spleen is very large, very soft, and easily reduced to a liquid pulp composed in great measure of blood.

In the thickness of the walls of the stomach may be seen bullæ due to gas, the muscular coat is blackish, the mucous membrane soft and easily torn off with the finger-nail.

The crypts of the small intestine are swollen and turgescent, and are milky-white as in carnivora during the act of digestion.

The remaining abdominal viscera are of normal appearance.

CASE XIX. *White wine taken in the morning, excess in brandy and absinthe.—First attack of alcoholic delirium occurring two years back.—Epileptic fit.—Second attack of delirium tremens with fever. Temp. 40° C. (104° F.) on admission.—Dyschromatopsy.—Anæsthesia, most marked on the right side.—Cure of the acute symptoms.—Persistance of the symptoms of chronic intoxication.*

C——, æt. 43 years, a dyer, the son of a tuberculous mother and of a drunken father who tried to hang himself, has for a long time given way to the habit of drinking. But for the last four years his excesses have been more numerous, he is addicted to taking brandy and absinthe, and drinks white wine in the morning on an empty stomach. He is changed in character, has become gloomy, suspicious, jealous, scolding his wife without any reason; his sleep is often disturbed by dreams and nightmares; he has lost his appetite and is troubled in a morning with vomiting of phlegm, white or green in colour, and sometimes streaked with blood.

Two years ago, in his first fit of alcoholic delirium, he heard crying, and singing; he saw lovers approaching his wife; he saw precipices, ships on the sea, rats, and spiders. At this time, according to his wife's statement, he trembled in all his limbs.

From this epoch he retained ideas of jealousy which betrayed themselves in insults and threats and sometimes in violence.

Fifteen days ago he entirely lost his sleep, he heard persons insulting him, saw rabbits, birds, men armed with muskets; and fancied that persons were cutting holes in the wall to shoot him with revolvers.

In order to escape his enemies he had carefully put a hamper in his place in the bed, upon which henceforward were to fall the bullets intended for him. On the third of June, eight days ago, he was suddenly seized with a convulsive attack, with loss of consciousness, falling, biting the tongue and having involuntary evacuations.

He came into the Asylum on the 7th of June. He is incessantly moving, looking on every side, throwing his hands about in all directions; he is in a condition of profound terror, utters broken sentences, sighs, is abusive, alternately threatens and supplicates, and in the midst of this disorder gives us to understand that he is condemned to death and is to be executed, that he sees his wife in another man's arms. Moreover he sees animals which he drives away with his hands and feet.

His features are changed, his eyes injected and bleared, his face covered with a clammy sweat, the lips crusted, tongue dry, the arms and legs trembling, together with muscular quiverings which extend at times to the body and the face. The skin is warm, the rectal temperature taken at 5 o'clock p.m. is 104° F., the pulse is hard, and small, 104 per minute.

The urine contains neither albumen nor sugar.

Treatment. Milk, broth, water mixed with wine and sugar.

He is very restless at night; the hallucinations are incessant; the perspiration is so profuse that two changes of linen have been required.

June 8th, morning. There is a little less restlessness, but the hallucinations are incessant, he sees fleas and flies on his arms and legs, hears threats and insults, &c.

The general tremor is less, but the trembling is still very marked in the arms and legs. The eyes have a dark circle round them, the sclerotics are injected and slightly yellowish, the tongue is moister. T.R. 38·6° C. (101·5° F.) Pulse 96.

The same treatment : three plates of soup.

Evening. Delirium the same; trembling of the arms and legs; shocks and quiverings in the muscles of the limbs and occasionally in the face and back. Pulse 100. T.R. 39° C. (102·2° F.)

June 9th. The hallucinations have persisted all night; this morning the patient is calmer; he says that people want to poison

K 2

FIG. 4.—C., ÆT. 43. CASE XIX.

Febrile Delirium Tremens.

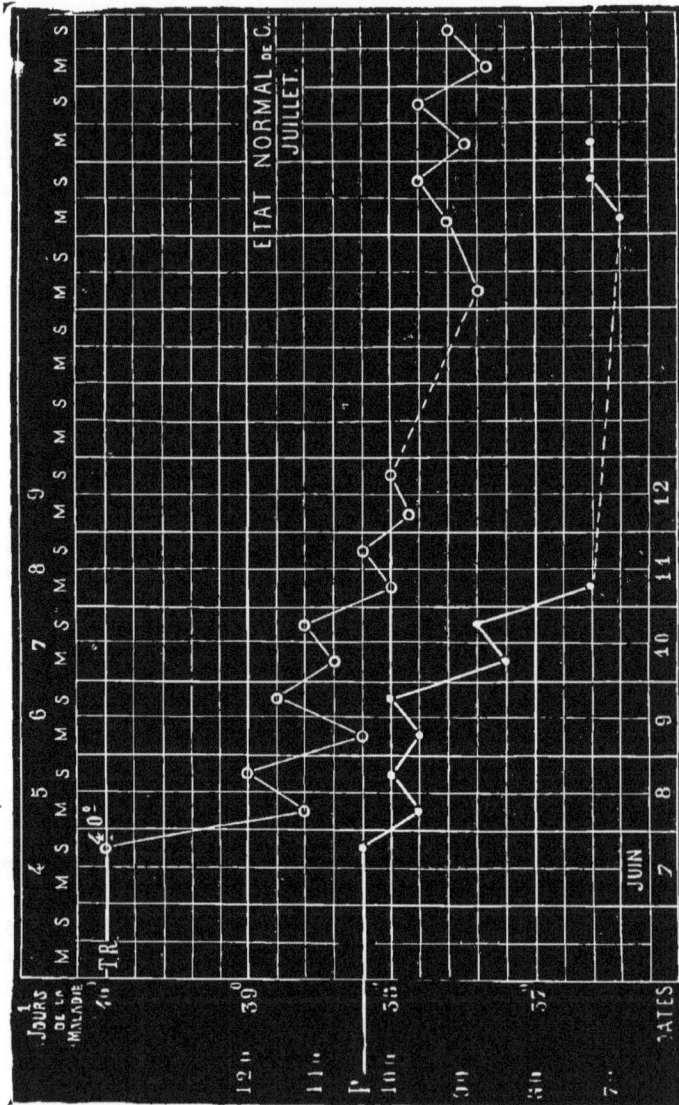

For explanation of the terms, see figs. 1 and 2, pp. 113 and 120.

him, men walk around his bed, he has seen his funeral, it was said that he was dead. People mock him and jeer at him about his wife's conduct. Headache, dizziness, trembling of the arms and legs.

T.R. 38·2° C. (100·8° F.) Pulse 96.

Same treatment; coffee with two ounces of sugar; a chop, 3 dishes of soup.

Evening. Pulse 100. T.R. 38·8° C. (101·8° F.)

June 10th. Slept 5 hours in the night, hallucinations persistent; in the morning he answers quite accurately, he believes in the reality of everything which has been the subject of his delirium. He still hears some insults. Trembling of the hands. T.R. 38·4° C. (101·1° F.) Pulse 84. Same treatment; a bath.

Evening. Pulse 84. T.R. 38·6° C. (101·5 F.)

June 11th. He has slept part of the night, but the hallucinations continue. In the daytime he is calmer, but he still believes that he is being persecuted, he is convinced that his wife deceives him, that persons bear him ill-will and try to injure him. He has a whitish mist before his eyes which obscures objects, he has difficulty in distinguishing colours; red he calls chestnut; but he is absent and responds but little to this mode of examination.

Pulse 70. T.R. 38° C. (100·4. F.)

Treatment; a bottle of lemonade, a dose of infusion of hops; coffee with two ounces of sugar.

Evening. Pulse 68. T.R. 38·2° C. (100·8°. F.)

June 12th. Sleep disturbed by night-mare; no hallucinations in the daytime, but ideas of persecution. A whitish mist before both eyes. The patient has suffered from trichiasis for a long time, and the left cornea is slightly dim over the outer half; the right cornea has the normal transparency; the fundus oculi examined with the ophthalmoscope shews no change of any kind, neither peri-papillary nor perivascular infiltration, nor obstruction of the venous circulation.

Chromatoscopy of the retina with the chromatic scale of Dr. Galezowski gives the following results to-day; blue (No. 10) appears violet to the patient, green appears grey, greenish blue is dark grey or slate grey, red is a brownish mahogany colour; on a second experiment, red is chestnut, rose colour is grey. Acuteness of vision is also diminished. C—— reads letters of one millimetre in diameter with difficulty.

Taste is blunted, a certain time is required to distinguish sugar and salt when placed upon the tongue. The sense of smell appears to be normal; orange flower water, camphor and essence of mint, are recognised quickly enough. The cutaneous sensibility appears to be dull over all parts of the body.

Headache, buzzing in the ears, cramps in the fingers, weakness in the legs.

T.R. 38° C. (100·4° F.)   Pulse regular, 76.

Treatment; a bottle of lemonade with wine, a basin of infusion of hop, sugared; two ounces of quinine wine.   A bath.

June 16th.   His nights are better, but some hallucinations still remain; he hears his name called, he is insulted, his wife is called for.   In the daytime he is calm, speaks of his hallucinations as imaginary things; but there is, says he, some truth in the midst of it all; his notions of persecution are no better.

The sensibility to temperature appears to be normal; hot and cold bodies are perfectly well recognised over all parts of the body; there is analgesia, but the pricks of a pin are tolerably well felt over the arms and legs.   The tactile sensibility is deadened, but by testing it with the æsthesiometer we obtain much more precise results, which differ on the two sides.   On applying the two points of the compass, on the anterior surface of the right leg, in the direction of the axis of the limb, the distance between the two points has to be carried to 11 centimetres for the patient to appreciate distinctly the double contact.   If the distance is less than 9 cent. the sensation of only one point is obtained; moreover if one pricks the skin with two pins at the same time separated by a distance of 8 centimetres the patient only feels a single prick, at the distance of 10 cent. he feels two.   On the other hand, if one asks him to point out the seat of the pricks with his eyes shut, he is wrong nearly every time; when he feels only one prick, although two pins are used, he indicates as the seat of the prick a point intermediate between the two pins, but situated ordinarily on the line which joins the two.   Sometimes the finger is placed nearer to one of the pricks than the other, sometimes the upper, sometimes the lower, without any rule.   When both pricks are felt at once the points which are pricked are not indicated any more exactly, the finger is placed sometimes above, sometimes below, but, as a rule between the two pricks.

On the anterior surface of the left leg, the sensation of two points is obtained at the distance of 7 cent., and the sensation of one point only when two are applied begins when the points of the compasses are 6 cent. apart.   On pricking with two pins 6 cent. apart the feeling of one prick only, is similarly obtained, whilst at 7 cent. apart the sensation of two pricks begins to be felt.

When the examination is made in a direction at right angles

to the axis of the limb, the distances are much less. For example, on the right leg when the two points are 5 cent. apart and placed transversely (and not in the axial line) the sensation of two points is very distinct. When 4 cent. apart it feels as if there were only one point. On the left leg two points are felt at an interval of 4 cent., but at 3 cent. the sensation is of only one. A similar result is found on examining the arms, the anæsthesia being the more marked in the right. On applying the compasses in the direction of the axis of the limb to the anterior surface of the right fore-arm an interval of 9 cent. gives a distinct sensation of two points; an interval of $7\frac{1}{2}$ cent. gives the sensation of only one point. In the left arm an interval of $6\frac{1}{2}$ cent. gives the sensation of two points; at an interval of $5\frac{1}{2}$ cent. only one point is felt. On applying the points transversely, the sensation of two points in the right arm is obtained at 4 centimetres, and of one only at 3 cent.; in the left arm two points are felt at $2\frac{1}{2}$ centimetres and one only at 1 centimetre 7 millimetres.

In order to shew these various results more definitely, we have represented the different distances required to feel in the same region the sensation of one or two points, by parallel lines of a length equal to the interval between the two points.

C—— suffers from heaviness of the head almost all day, from dizziness and occasionally from cramps in the arms and legs, the latter are not apparently any more marked on the right than the left side; the muscular weakness is general, but is, however, a little more pronounced on the right side.

The patient by hopping on his left foot can manage a distance of about sixteen yards quite easily; with the right he hops more awkwardly and is obliged to give in before reaching the terminus. With the dynamometer the difference is very small between the right and left hand, the former gives 28° and the latter 29°. T.R. 37·8° C. (100° F.) Pulse 76.

Treatment: A bottle of lemonade with wine, a draught of infusion of hop sweetened with sugar. Quinine wine two ounces.

June 20th. C—— sleeps well, he walks about in the daytime and seems to have no hallucinations, but he is absorbed, and knits his eyebrows when speaking of his wife and considers his delirious ideas to be true and exact.

Heaviness of head, dizziness, a feeling of weakness at times, but never any vertigo. Treatment the same.

June 30th. The improvement continues. C—— receives his wife's visits with more pleasure, he dwells upon all that has

COMPARATIVE TABLE OF THE DEGREE OF SENSIBILITY BETWEEN THE TWO HALVES OF THE BODY IN THE PATIENT, IN CASE XIX, C——, Æt. 45 years, a dyer. Admitted into St. Anne, June 7th, 1872.

TESTING OF THE SENSIBILITY IN THE AXIAL DIRECTION OF THE LIMB.

(Anterior surface of the Leg.)

R. LEG.   11 cent. Sensation of two points.

9 cent. Sensation of one point.

L. LEG.   7½ cent. Sensation of two points.

6 cent. Sensation of only one.

EXAMINATION MADE TRANSVERSELY TO THE AXIAL LINE.

(Anterior surface of the Leg.)

R. LEG.   5 cent. Two points felt.

4 cent. One point only felt.

L. LEG.   4 cent. Two points felt.

3 cent. 3 millimetres. One point felt.

Examination of the Sensibility in the Axial direction of the Limb.

(Anterior surface of the Fore-arm.)

R. ARM.
9 cent.　Sensation of two points.
7½ cent.　Sensation of only one point.

L. ARM.
6½ cent.　Sensation of two points.
5½ cent.　Sensation of one point.

Examination made Transversely to the Axial Line.

(Anterior surface of Fore-arm.)

R. ARM.
6 cent.　Two points felt.
5 cent.　One point felt.

L. ARM.
2½ cent.　Two points felt.
1. c. 7 millimetres.　One point felt.

passed and evidently makes an effort to find out the truth : nevertheless his mind still retains groundless misgivings and delirious
ideas which become apparent on dealing a little more closely with
him.

His head continues to feel heavy and occasionally his arms and
legs feel numb. Examination of the tactile sensibility shews a
less degree of anæsthesia, but the general results remain the same,
the anæsthesia being a little more marked on the right side.

On the 15th. of July, being in a good way towards the cure of
the acute symptoms C—— was sent into the country.

This case deserves our attention on several
grounds. We see an attack of epilepsy occurring
at the onset of the second attack of alcoholic delirium, at a period when the trembling had not
as yet attained its highest degree of intensity.
There is then, as we have already pointed out, no
relation between the acme of the trembling and the
convulsive attacks ; these two orders of symptoms
belong to different causes. The attack of epilepsy
is due in this case to absinthe, perhaps also to the
white wine taken in a morning.

The fever, which we were enabled to estimate accurately by the aid of the thermometer, gave an especial gravity to these attacks of delirium tremens,
which the other symptoms, equally intense, did not
contradict. The tremulousness of the whole muscular system, the changed countenance, the injection
of the eyes, and the general aspect of the patient
made one fear a fatal issue, and with good reason ; happily, however, from the second day, the
fall of temperature and the slight improvement in
the other symptoms indicated a favourable tendency, which was not belied by the result. We
were obliged to keep up the strength of the patient, and as soon as he could take eggs and a
little meat we did not hesitate to give them.

The examination of the sensibility of the retina with respect to the perception of colour likewise furnished very distinct results, and finally, the testing of the tactile sensibility by Weber's method enabled us to point out with more precision than by other means a predominance of the anæsthesia on the right side.

There are still many desiderata in this kind of research; but, nevertheless, the knowledge already obtained on diseases of the cord, and some facts also in relation to cerebral pathology permit the hope that, little by little, light will be thrown on this difficult study.

CASE XX. *Abstract.* Victor A., æt. 30 years, a stone-engraver, addicted for several years to alcoholic drinks, was admitted into St. Anne on the third day of an attack of febrile delirium tremens. On admission, he is restless and full of hallucinations (he sees barricades, hears shells fall; he sees robbers and animals; answers his comrades, &c., &c.); he has considerable trembling of the limbs and muscular quiverings through the whole body. At the end of four days the delirium is better, the hallucinations only appear at night and are less numerous; the trembling has diminished, but the quiverings persist in the muscles of the arms, trunk and thighs.

The temperature, after having reached 40·6° C. (105° F.), falls, and for twelve days oscillates about 39° C. (102·2° F.), afterwards falls to 38° C. (100·4° F.) and even lower, then becomes normal all at once. The defervescence is followed by the diminution and disappearance of all physical and intellectual symptoms, and the patient leaves the asylum cured.

I have chosen the preceding cases of febrile delirium tremens terminating in recovery, in order to shew two different modes of the decline of the fever. The tracing in case 19, patient C——— shews the most common course of the fever. The line falls from 40° C. to 39° C., it then presents oscillations, falling in three to four days from 39° C. to

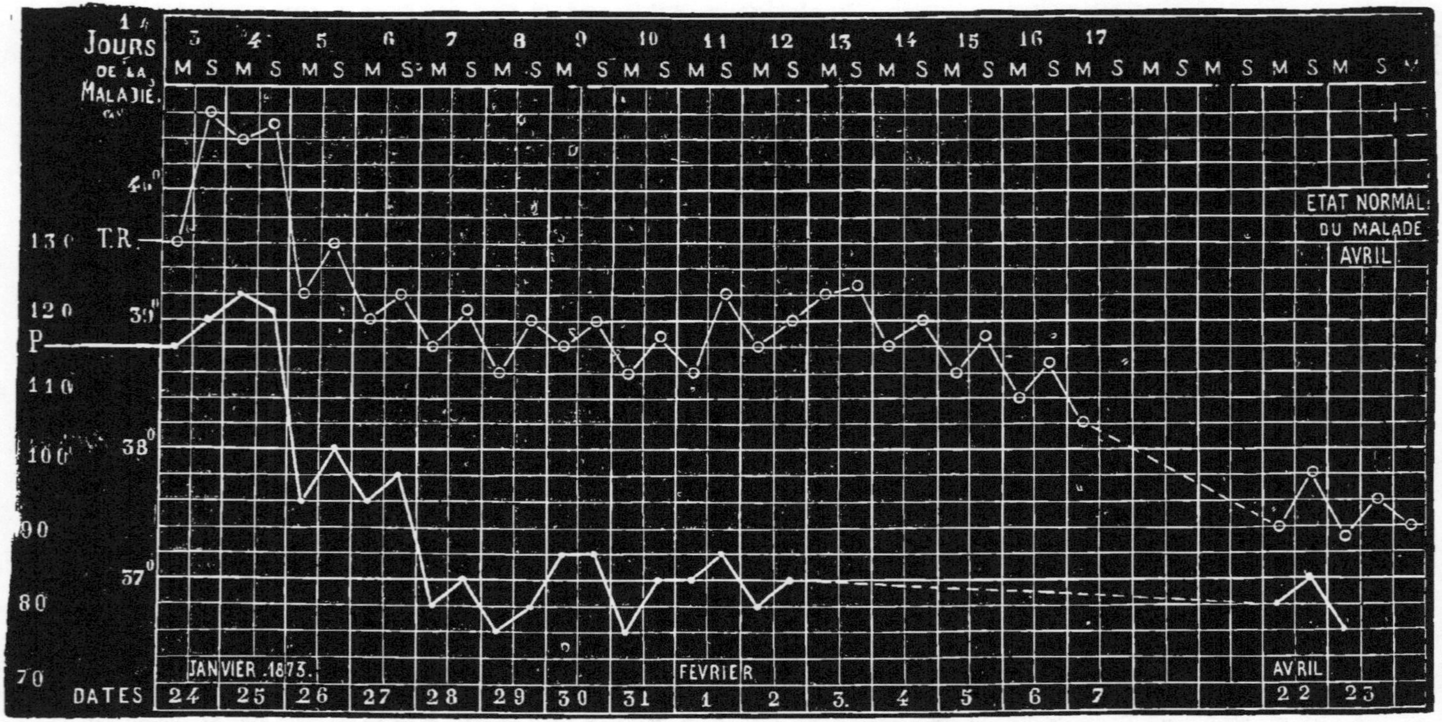

Fig. 5.—Victor A——, Æt. 30. Case xx.

38° C., and finally reaches the normal temperature of 37·4° to 37·6° C. In Victor A——, case 20, the tracing presents a rarer type, it shews irregular oscillations and then a slight transient rise of temperature at the eleventh day of the disease, after several days of descent.

The other symptoms generally follow the march of the temperature and amend in proportion as it declines towards the normal condition.

ARTICLE II.

Differential Diagnosis of Febrile Delirium Tremens and Delirium Tremens associated with intercurrent diseases and with injuries.

Febrile delirium tremens, free from all intercurrent disease, and principally characterized by the peculiar trembling of which we have spoken, is radically distinct from delirium tremens associated with intercurrent affections and injuries.

(*a.*) Febrile delirium tremens occurs almost always after recent and numerous drinking-excesses; whilst the delirium tremens associated with intercurrent affections breaks out in the chronic alcoholic, without fresh excesses, and only on the occurrence of the disease which has accidentally attacked him.

(*b.*) In febrile delirium tremens the fever is an essential element of the disease, and depends only upon the disease itself; in the other forms of delirium tremens the fever has various sources, from pneumonia, pericarditis, erysipelas, &c., even to injuries of accidental or therapeutic origin.

(*c.*) In both classes of cases we must take into account the degree of chronic alcoholism which the person has reached; but, whilst in the former the seriousness of the case depends entirely on the intensity of the acute attack, in the latter it depends on the super-added malady.

(d.) In the pathological combination of an intercurrent affection and alcoholism, the alcoholic symptoms accidentally developed present very various degrees of intensity; not only does the delirium sometimes betray itself simply by nocturnal hallucinations and nightmare, but the motor disturbances are often limited to a more or less extensive trembling of the hands, or of the arms and legs, or of the arms, legs and face, and are never accompanied by the alarming tremor of all the muscles of the body which occurs in febrile delirium tremens.

(e.) In febrile delirium tremens, with the exception of the hyperæmia which sometimes leads to hæmorrhage and thus bears witness to the violent irritation of the nervous centres, we find very little in the various systems at the *post-mortem* but the changes due to chronic alcoholism (steatosis and sclerosis) in a more or less advanced state; the other form of delirium tremens, besides the lesions due to chronic alcoholism, shews also the peculiar lesions caused by the intercurrent disease, which are often very extensive and are the principal cause of the fatal issue.

(f.) Let us add that the subjects of febrile delirium tremens, on account of the violence of the delirium, are generally placed in asylums, whilst patients attacked with delirium tremens in connection with intercurrent diseases are more generally treated in ordinary hospitals.

ARTICLE III.

Treatment of Febrile delirium tremens.

The differential diagnosis between febrile delirium tremens and delirium tremens associated with intercurrent affections, was indispensably necessary for the purpose of considering the treatment; and now we can easily see that this will not be the same in the two forms of delirium.

Moreover, delirium tremens associated with intercurrent diseases, since it appears under the most varied aspects, furnishes in each of its form's the grounds of special therapeutic indications; let us hear how M. Gubler expresses himself on this subject: "Delirium tremens is not an entity comparable to a created species, always sufficiently like itself for each case to represent the unit, or in other words, the diseased individual. On the contrary, it is an eminently variable combination of symptoms, varying in the time, the form, and the intensity of the symptoms according to the peculiar conditions of the patient and his surroundings. In the midst of this symptomatic polymorphism, the original identity in etiology loses, so to speak, all its value in the eyes of the practitioner." In this sense we must admit the truth of the adage : " There are no diseases ; there are only sick people."*

It would be in vain therefore to try to lay down therapeutic rules applicable to every case.

If we find, indeed, that in all these patients one part of the danger arises from the common ground which belongs to them, that is to say, a more or

---

* Gubler. *Bullet. de l'Acad. de Méd.* Vol. **xxxv.** Meeting of Dec. 20, 1870.

less advanced stage of chronic intoxication, it is no less true that with regard to the other element, the intercurrent disease, they differ in the most complete manner. But it is not so in febrile delirium tremens, this ordinarily presents itself under a type sufficiently uniform for us to discover certain general indications which are applicable to the majority of cases if not to all.

Almost all patients suffering from febrile delirium tremens present a state of violent agitation, at least in the early part of the attack; from this fact one indication arises as a matter of necessity, viz., *to protect the patient against himself and to prevent his injuring those who are about him.* The way in which this is to be effected is not a matter of indifference; we shall return to it bye-and-bye.

Again, we know that for more than three days after giving up excesses, alcohol still remains in the tissues, it is even probable that some remains till the fourth and perhaps even to the fifth day, and up to the present time we have been able, once at least, to ascertain that there was none remaining at the seventh day. Be this as it may, the danger is especially great at the beginning of the attack, since deaths are most frequent from the first to the fourth day, that is to say at the time when the system is under the effect of this toxic impregnation, the influence of which must be the more deadly as it is exercised upon elements of extreme sensibility, upon nerve-cells which are already jaded and exhausted by the enormous activity to which the poison has condemned them. It is therefore of the first importance to rid the system of this dangerous enemy. And lastly, such a violent and excessive degree of excitement is not without after-effects. To this period of over-action there soon succeeds a depression and prostration which is not the least danger of this stage of the disease; hence arises a third ground for treatment which must be taken into account.

To sum up: the efforts of the practitioner should have for their aim—

(1). To protect the patient.

(2). To promote the elimination of the poison.

(3). To maintain and restore the strength.

I. To protect the patient against himself and to prevent his injuring those around him.

This indication, which we should be inclined to consider the most important, is very difficult to put into execution; for, on account of the special conditions in which the patient affected with febrile delirium tremens is found, it is necessary on the one hand to avoid preventing all movement and fixing him to the bed, which favours the stasis of blood and the congestions, which are so common in this disease, and which also obstructs the expulsion of the poison; and, on the other hand, to avoid the use of means for quieting him which would prove a serious obstacle to the full play of the chest walls, either by preventing the expansion of the base of the chest, or by compressing the air-passages in the cervical region.

Hitherto, the practice generally adopted in France, not only in asylums but also in hospitals, has been to apply the strait-jacket to, and to fasten upon his bed, every patient whose violence was feared either for himself or for others. There is no need for us to describe the straitwaistcoat,* an apparatus which is well known and whose dangers and in-

* Rouhier, *De la Camisole ou gilet de force.* "*On the camisole or strait waistcoat.*" (Thèse de Paris. 1871.)

conveniences have furnished matter for numerous
dissertations; but it will be well for us to recall the
sad consequences which result from the application
of this instrument in combination with fixture to the
bed : here is what we have already said on this subject in the article " *Camisole*" in the Encyclopædic
Dictionary of the medical sciences; " The patient is
stretched on his back ; thanks to the camisole, there
is already considerable restraint upon the base of the
thorax, and the false ribs and the two or three lowest
ribs are in great measure prevented from moving.
In order to secure the upper part of the body, cords
are passed through the two rings situated at the
level of the sub-clavicular regions, and these are fastened below and behind to the head of the bed; further, in order to prevent the patient from slipping,
the pillow is pushed between the cords which pass
under, and the head which is applied to and rests
upon it, so that all the anterior and upper part of
the strait-jacket is tightly stretched over the corresponding parts of the thorax which it fixes.   The
chest being thus restricted at the base and apex, and
only an insufficient play of the thoracic walls being
possible, the movements of the diaphragm are exaggerated, and thus an abdominal respiration is produced which to a certain extent makes up for the defi-
ciency of thoracic respiration.   But the patient, feeling himself firmly held at the upper part of the body,
becomes so much the more fidgety, and throws his
legs in all directions ; the latter are then attacked,
and the feet, bound in fetters, are fastened to the
lower end of the bed.   The effect of this is to make
the muscles of the abdominal wall tense and so
hinder the descent of the diaphragm which is so
essential.   Lastly, strong sheets, thrown across the

belly and the knees, are fastened under each side of the bed; this last precaution completes the list of means so well combined to produce gradual asphyxia, as the examination of the viscera at the *post-mortem* demonstrates. Let the patient begin to struggle or make efforts while in this condition, and the face becomes rapidly injected, the eyes glistening, the turgid jugulars empty with difficulty, and the swollen neck is strangled against the rigid edge of the strait-waistcoat. Not to mention the cases in which the unfortunate patient in his endeavours to get free finishes by slipping over one side of the bed; and the larynx, pushed by the weight of the whole body against the border of the camisole becomes the seat of serious lesions."

We might here relate some autopsies of alcoholics in which the condition of the lungs would forcibly shew the mischievous action of the strait-waistcoat, but as in alcoholics pulmonary congestion may occur without the aid of this means of restraint, these cases might leave doubts in the mind, so it may not be without advantage to see what occurs under similar circumstances in simple cases, for instance in a maniac.

Case XXI. *An attack of mania without fever—The patient fixed and prevented from moving in bed.—Asphyxia.—Death.—Pulmonary apoplexy.*

Jean J——, aged 47 years, a coach painter, admitted into an asylum on the 12th of November.

This vigorous man, habitually of good health, had felt jaded, slept badly and had been restless for about three weeks. For six days past he had been loquacious, incoherent, and excited, and on admission shewed that he was a prey to a violent attack of mania with hallucinations, his ideas and actions confused, loquacity, cries, songs, and incoherence, but *no fever.*

He was allowed to be free for five days, the maniacal attack continuing with the same intensity. From the 17th to the 23rd of November he was kept fastened in bed *a part of the day-time;* the restlessness continued, but the appetite was maintained, and there was no fever.

The patient was kept *constantly in bed* from the 23rd to the 26th of November, he became much oppressed on the 25th, breathed with the greatest difficulty and died on the morning of the 26th, without any symptoms except laboured breathing, injection and later on, puffiness and cyanosis of the face.

Autopsy. The lungs, which are of a deep red colour, present blackish-red, irregular, slightly prominent patches on the surface, from an half inch to one and a quarter inches in width; on pricking these blackish spots with the point of a bistoury, liquid blood flows from them. On cutting into the lungs one sees on both sides in the thickness of the organ, but especially at the back and towards the base, little blackish islets from the size of a hazel nut to that of an almond, from which there escapes on pressure a liquid of a deep red colour (patches of pulmonary apoplexy).

The heart is normal, the orifices healthy, the right cavities enclose some blackish clots.

The cerebral meninges are thin, but very injected; the pia mater which is infiltrated with blood at the convexities of the hemispheres is easily separated everywhere; the small veins which border on the sinuses are turgid and blackish. All the brain is injected and the grey portions shew a slight rose-coloured tint. The lateral ventricles contain from 30 to 40 grammes (about 1 ounce) of bloody serosity.

The other organs shew no changes.

Seeing that the camisole is capable of producing such serious symptoms in ordinary patients, what may it not do in alcoholized patients who are so strongly predisposed to chest complications? Still, however, these patients cannot be left at liberty at the moment of their maniacal paroxysms; we must use a means of restraint which anticipates these accidents and without embarrassing the bodily functions restricts the excessive agitation of the alcoholic maniac. M. Bouchereau and myself in our

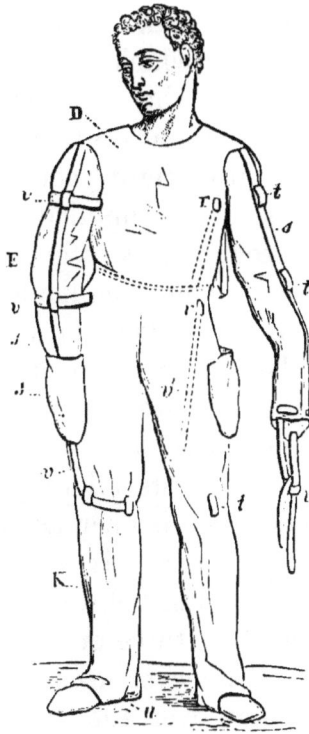

FIG. 6.

wards at St. Anne make use of a dress which appears to fulfil these indications.

"The *Maillot* (lit. long-clothes or tights) is made of canvas, and forms as it were the lining of a garment made of cloth for winter and of light material for summer. It fits pretty exactly over the surface of the body without exercising pressure at any point. The body of it is provided with four eyelets (*r*) on each side, placed two and two at the same height for passing cords through (*v*); the latter surround the arms and are fastened behind the patient; they are sewn inside and remain concealed when not in use (v'). The sleeve (E) has two small loops (t) on the outer side, which to make them more secure are fastened to a band (s) which is sewn to the whole length of the sleeve; the cords (v) pass through these loops, which by preventing the cords from slipping, render any tight constriction useless. The end of the sleeve is supplied with eyelets and when these are brought close together by a cord a *cul-de-sac* is made in which the hand is lodged; this is then placed in the pocket (I) which is open at the bottom for the passage of cords (v) which pass round the trouser and are fastened behind; loops (t) serve to support these, and the strap under the foot (u) in its turn resists all displacement. The back of the dress which is open as far as the lower part about the perineum, is fastened as low as the waist by a cord passed through eyelets; the remainder is fastened by three or four buttons which are easily undone whenever necessity requires it."

A petticoat is put over the dress in women, in order to avoid the singularity which such a garment would have on them. When the dress is well made and well applied it holds the patient as fast as the strait-jacket, without having the inconvenience of restricting the breathing or the circulation. When the patient is struggling, the position of the arms on the sides of the body, the reverse of what occurs with the camisole, favours the separation of the walls of the dress and renders the respiratory movements so much the easier. Again, the numerous points of support which the fastenings have, divide the pressure over large surfaces and so present very great advantages not only over the camisole, but also over such apparatus as Haslam's girth with lateral bracelets, which is apparently simpler, but is more objectionable when applied. In short the dress constitutes a complete garment; warm in winter, cool enough in summer, but covering the whole surface of the body, and thus guarding against the serious accidents which the action of cold produces in patients who, in spite of careful watching at night, expose themselves, get out of bed and remain lying on the ground for a longer or shorter time.

The patient thus guarded by the dress may without inconvenience in the majority of the cases be left in a room with other patients in the day-time, or better still, in the court when the weather is fine; when night comes, the patient, still wearing the dress, which has the great advantage of maintaining a uniform temperature over the whole surface of the body, is placed in a mattressed cell with a double layer of straw-matting on the floor, and having a mattress and bolster in one corner. This is sufficient; but it must not be forgotten that before being placed in the cell he must have been kept for a sufficient time on a close-stool to empty his bladder and rectum.

Thus installed, our alcoholic may struggle, cry, dash himself against the walls or throw himself on the ground, and no accident need be feared, and the more his movements are increased, the more active will be the circulation, the perspiration profuse, the respirations deep and numerous; all these being physiological conditions which as we can understand are eminently favourable to the restoration to health. Under these conditions, when the patient, worn out by his incessant movements and by his excessive restlessness lies down on the mattress, we soon see him plunged in a profound calm sleep in

which the respiration is free and the whole surface
of the body is covered with a moisture produced
slowly, uniformly, and everywhere in the same
manner.   Then in the morning, a bath for half an
hour or an hour, by freeing the skin of all the ex-
cretory products disposes it to pursue the necessary crisis in a favourable manner.

II. The second indication, viz., to eliminate the
poison, becomes very easy of fulfilment as a result
of the preceding conditions.   The patient perspires
abundantly and breathes easily, we have already
therefore two large channels of exit open to the
poison; diluent drinks of slightly aperient or diuretic character, but above all given copiously, whilst
satisfying the intense thirst of the patients, open the
third channel of elimination, and the kidneys, coming to the help of the skin and the lungs, place the
alcoholic in the condition most favourable to the
expulsion of the toxic agent and also to the regular
course of all the functions.

III. To this period of exaltation and of extreme
agitation at the onset, and to this tremulousness of
the whole body there succeeds a collapse, a profound depression which may become a serious peril
and threaten life; we must therefore keep up the
strength.   We have recourse to milk, broth, chocolate, coffee, to wine diluted with water, to quinine,
the liquid extract being preferred, in doses of 2 to 6
grammes either in a draught with syrup of orange
peel, or as an electuary in combination with con-
fection of roses; we must nourish the patient as
rapidly as possible, giving eggs and meat and especially the gravy of meat whenever the administra-
tive authorities do not oppose it.

The Academy of Medicine, when discussing the

gravity of traumatic lesions in alcoholics, had chiefly in view delirium tremens associated with intercurrent diseases and injuries; and in that learned
discussion it was generally agreed that but little
value was to be attached to bleeding, tartar emetic
and digitalis. On the contrary, opium in moderate
doses, wine, coffee, and tonics generally, obtained
the most support; but, as we have already stated,
each particular case presents itself with its special
indications, and one sees oneself obliged every day
in practice, to adapt the treatment to the exigencies
of the individual case.

We will add a word in reference to chloral hydrate, the efficiency of which in delirium tremens
has been so vaunted that certain physicians would
make it almost the specific for this disease. Chloral certainly is found applicable in the treatment
of mental diseases, and it can ensure a very favourable sleep in mania, and in melancholia with restlessness and insomnia, but if we recall the physiological properties of this drug it will be sufficient
to make us reject its use in the acute stages of a
poisoning which of itself compromises in so energetic a manner both circulation and respiration.
Moreover, on looking through most of the cases of
delirium tremens cured by hydrate of chloral, one
soon discovers that these cures were not in reality
of cases of febrile delirium tremens, but rather of
simple alcoholic delirium.

CHAPTER IV.

CHRONIC ALCOHOLISM.

ARTICLE I.

Intellectual Symptoms.

WHEN the intoxication is prolonged, whether this occurs in a slow and progressive manner, or after several relapses with acute symptoms, there often remain, as M. Falret has pointed out,[*] some delirious ideas, hypochondriacal preoccupations, illusions and sometimes even hallucinations, which are a reflection of the sensory and intellectual disturbances of the earlier stages, but which have neither their activity nor their acuteness; the outbreaks of delirium are milder, the excitement is less sudden and less noisy, without, however, completely disappearing. Chronic alcoholics, in fact, are not exempt from melancholic delirium and suicidal ideas, but these vaguely expressed disturbances are, I repeat, nothing but a rough sketch of the symptoms which occur at the commencement of the poisoning in the stage of acute alcoholism. And, lastly, in some patients there come on at irregular intervals, with or without fresh excesses, turbulence and a state of semi-maniacal excitement in which the person disorders or tears his clothes, and runs about hither and thither, in an automatic manner.

In this chronic stage of intoxication, a new element comes into play; the condition of the subject is modified; the seed sown is the same, but the soil is

---

[*] J. Falret, *Recherches sur la folie paralytique et les diverses paralysies générales,* 1853, p. 119.

different, the alcohol has already made another person of the man. It is not in vain that the alcoholic has subjected himself to the constant action of the poison, the latter has built up its work little by little. At length, we no longer see only functional disturbances, and transitory changes, leaving in their train only a slight malaise; a more profound action has been produced, nutrition is altered throughout all the organs, systems and tissues; the cell itself lives with another life. Under the influence of alcohol a double morbid process has developed itself; the system as a whole has been struck, as it is said, with premature old age, and has undergone fatty degeneration; but the fatty degeneration is not alone, together with it there occurs a tendency to diffuse chronic irritations, and this double process, (sclerosis and steatosis) becomes the characteristic of chronic alcoholism. And according to the predominance of the one or the other of these lesions in the nervous centres, do we see the chronic alcoholic progressing either towards dementia (steatosis and atheroma), or towards general paralysis (diffuse interstitial sclerosis.)

These organic changes appear, moreover, with the same characters in the muscles and the glands; and as regards the liver in particular, we know that either cirrhosis or fatty degeneration is the concomitant of chronic alcoholism.*

But to return to the intellectual phenomena. We find in chronic alcoholism, that the memory is weakened, the judgment less sure, and incapable of discernment, the imagination extinct, the faculty of association of ideas greatly diminished, and no

* Lancereaux, article *Alcoolisme. Dict. encycl. des sciences médicales.*

longer preventing incoherence, and lastly, that the moral sense is greatly blunted.

Apathetic, indifferent, stupefied, the chronic alcoholic bestows no attention on his person, he takes no care of his family, he is lowered in all his intellectual, moral, and social faculties, and finds himself yielded defenceless to the caprices of his instinctive appetites. It is at this moment that we may repeat the old proverb, " *Who has drunk will drink ;*" the patient is driven to it in different ways, and reason affords no counterpoise sufficient to prevent him. Excesses in drinking have become a habit; incited, moreover, by his hypochondriacal tendencies, the habitual drunkard renews his excesses, in order to get a remedy for his woes, to regain his strength, and to check the uncomfortable retching in the morning. The results of this medication speedily appear; the disquietude increases, and hence the vicious circle from which the unhappy sufferer is released only by falling into the most absolute dementia.

In the last stages, the intelligence becomes a void; insensibly all the delirious outbreaks disappear, the hypochondriacal ideas, and all the sensory disturbances, are by degrees effaced. Sometimes a maudlin sentimentality supervenes, as in cases of senile dementia and apoplexy, but all these symptoms are equally indicative of profound lesions.

ARTICLE II.

Physical Symptoms.

The motor and sensory physical disturbances accompanying chronic alcoholism are very variable, and necessarily so, since they depend upon two different morbid processes, steatosis (fatty degeneration) and atheroma of the vessels on the one hand, and sclerosis on the other, each of which is capable of giving rise to the most diverse local affections. As regards the nervous system, we may nearly always distinguish two sets of symptoms, the one set peculiar to the spinal cord, and the other excited by the cerebral lesions, and as in other affections of the nervous system, in general paralysis, for instance, these two classes of cerebral or spinal symptoms do not follow any regular order in their development.

In some cases the predominance of the spinal disorders would lead to the belief that the whole action of the poison was concentrated on the cord ; we may in fact, find, at a very early period, and even whilst the intellectual disturbances are but little marked, that there are pricking sensations in the skin, especially of the soles of the feet and of the hands, numbness and tingling, abnormal sensations of heat and cold, cramps, superficial or deep hyperæsthesia, muscular or cutaneous anæsthesia, weakness and trembling of the arms and legs, a sluggish condition of the contractile reservoirs (bladder and rectum—incontinence and retention), there exist, in a word, all the phenomena connected with diffuse myelitis to which M. Hallopeau has drawn attention.[*] More rarely it happens that the symptoms observed are those of a fasciculated sclerosis, limited to definite bundles of fibres of the cord, but even in these cases, it is exceptional to find them pure and unmixed ; for instance, together with the signs of sclerosis of the posterior columns, one finds associated the symptoms peculiar

[*] Hallopeau, *Etude sur les myélites chroniques diffuses* (*Arch. gén. de méd.*, of Sept., 1871 to Feb. 1872).

to diffuse sclerosis, as in the two cases of alcoholism ending in general paralysis, (Cases xxv, xxvi.)

But apart from the diffuse scleroses mentioned by authors, there are also the lesions depending upon fatty degeneration and atheroma of the vessels, with their numerous varied localizations in the grey and white matter. These lesions are less known than the various forms of sclerosis, or at least the relations which exist between them and the symptoms which they excite have not as yet been clearly determined. Here, we think, will be the second phase of the study of the pathological anatomy of the lesions of the cord in chronic alcoholism.

Passing on now to the cerebral disturbances, motor and sensory, we still find ourselves in the presence of the most complex set of symptoms, since steatosis and vascular atheroma on the one hand and sclerosis on the other, produce lesions throughout the whole extent of the brain. We do not know, in fact, what symptom there is which might not be associated with chronic alcoholism under one or other of these conditions. The whole encephalon may be affected in the greatest variety of ways, either partially or to a considerable extent, by diffuse or circumscribed lesions ; and these may result in attacks of local, sudden and instantaneous character, or of very slow development. Amongst all these symptoms, there are some which exist in the majority of chronic alcoholics, but which have nevertheless no special characters, and may be met with in other diseases. Headache, for example, with heaviness and a sense of weight in the head, hummings in the ears, attacks of dizziness and giddiness and sometimes of actual vertigo, are common symptoms of chronic alcoholism ; and they are sometimes accompanied by numbness or tingling in the arm or leg of one side, and by unilateral muscular weakness of short duration. In other cases, after a more or less prolonged malaise, persons affected with chronic alcoholism become pale, grow weak, fall or sit down and remain for one or several hours plunged in a semi-comatose or comatose condition, from which they often recover with paralysis of one arm, one leg, or one side of the body, which are of very variable intensity and duration, (Cases xxiii, xxx, &c.) At other times they fall, or else whilst lying down in consequence of headache, dizziness or vertigo which have in some measure fore-warned them, they are seized with slight convulsive twitchings of clonic character, and either limited to one or several members or generalized, and which are more or less rapid, more or less marked, and with or without grimaces of the face. These convulsions may recur five or six times in succession and return after some days or cease altogether. And we must observe that the patient usually retains a recollection of them, and that, sometimes, he does not completely lose consciousness; and that these attacks come on even during his stay in the asylum, and consequently, under hygienic conditions from which all causes of excitement are as a rule excluded.

These epileptiform attacks, which, as we know, depend upon lesions already produced in the brain, are not in any way comparable with the attacks of epilepsy which occur with the acute symptoms of alcoholism, and which depend for the most part upon recent excesses in absinthe; although in some cases, in persons who have been long addicted to this liquor, the attacks may persist for some time after the cessation from such excesses, (Case xii).

Magnus Huss, whose book is so remarkable in many respects, has not, in our opinion, sufficiently determined what part of the symptoms should be ascribed to the spinal cord in chronic alcoholism.* On the other hand, the division into six forms given by that author, does not group the facts in a really clinical manner. This is his division :

1. Prodomal form ;
2. Paralytic or paretic form ;
3. Anæsthetic ;
4. Hyperæsthetic ;
5. Convulsive ;
6. Epileptic.

When we come to analyze the cases carefully, we see that many of them might, with equal justice, belong to two or three of these forms. Thus the patient in case xxx might be classed either under the paralytic or the anæsthetic form ; the patient in case xix would come equally well under the convulsive form as under the anæsthetic or paralytic ; and the same with many other cases. We have thought it right to make these remarks with reference to a work justly esteemed, and whose value cannot be contested.

ARTICLE III.

Cases of Chronic Alcoholism with a tendency to Dementia.

In the three following cases we see chronic alcoholism tending to dementia. In case xxii the

---

* Magnus Huss, *Alcoholismus Chronicus*, Stockholm, 1852; German translation by Gerhardt van den Busch.

decay of the faculties has made especial progress after poisoning by phosphorus.

CASE XXII. *Of drunken habits; admitted into asylums eight times.—Alcoholic delirium of maniacal and melancholic forms.— Ideas of homicide and suicide.—Poisoning by phosphorous (hæmoptysis, purpura).—Impairment of intellect with ideas of persecution.— Tendency to dementia.*

Eugene H——, æt. 38 yrs., a sculptor, had already been under treatment four times in the course of a few years at the Bicêtre for alcoholic delirium when he was brought to the St. Anne Asylum on May 16th, 1867. H—— was accustomed to drink wine and brandy, he ate little, was tormented at night by painful hallucinations, which for some time past had persisted also in the day time and had given rise to notions of persecution. He imagined that his wife wished to get rid of him in order to marry again, that his companions fought shy of him, and wished to injure him; and that he was accused of robbery and wilful murder.

On admission he refused to answer questions, walked about in silence with his head cast down, turning about at times and looking in various directions. For a moment his features would be contracted, he would stop with an angry expression and then suddenly dash against the door, head foremost.

The hands were tremulous, the tongue white, pulse regular and compressible, pulsations 76.

Treatment. A bottle of seidlitz water; two bottles of citric lemonade; a bath.

The night passed without sleep. He got up, struck against the doors, cried out and occasionally used threatening language.

May 17th. Delirium the same; persistence of the hallucinations; behaviour the same. The patient ate and drank without difficulty.

Treatment. A bath. Two bottles of citric lemonade.

H—— slept four hours during the night; in the morning he appeared to be less absent, and answered in an abrupt manner a few words relating to the painful hallucinations. It was not until the fifth day that he related how people had a grudge against him, sought to injure him, threatened him and were constantly following him. He complained of headache and of numbness in the limbs.

From this time the improvement continued and the patient went out at the end of three months.

Once outside, he resumed his work and also his drunken habits. The hallucinations and delirium quickly reappeared, and H—— was arrested in the streets armed with a knife and chasing the passers by. He was brought back to the asylum on October 24th, 1867.

He was frightened, stooped, and tried to hide himself; they wanted to kill him, he said, persons threatened him, and fired guns at him. At the end of a few days the hallucinations were less frequent, and tended to disappear, but there remained some brain-disturbance, a little stupidity and ideas of persecution; convalescence was slowly re-established and H—— remained in the asylum until May 8th, 1868.

On his return home he rested some days before resuming his work; his wife stated that he was quiet and comparatively sober, that he had no more delirious ideas, and that everything went on very well for a fortnight; he had however attacks of dizziness, complained of his stomach and ate but little.

Towards the 20th of May, his sleep became disturbed, and the delirium returned. H—— saw himself pursued by policemen, he was to be guillotined; he was told that his wife would forsake him. He left his house, roved about the streets, sometimes threatened his wife and appeared to be a prey to a deeper and deeper restlessness.

On the morning of May 27th he swallowed in succession three half glassfuls of water in which he had allowed about half a pound of matches to macerate for some minutes (this statement was made by the patient afterwards, he having been alone at home at the time). During the day he did not go out, but drank of the infusion without telling his wife what he had done. He passed a bad night, the hallucinations were incessant, "they shot arrows at him, wanted to murder him, wished to arrest him". H—— was quiet for a little while in the morning and his son saw him swallow a fresh infusion of matches. During the day he was in a state of extreme excitement, he heard insults and threats on every side; he shut his wife up in a room, told her she must die with him, and tried to strangle her.

H—— came into the asylum on May 29th, he was full of hallucinations; he was accused of murder, and was about to be put to death; the police were seeking him; a copper was being prepared into which he was to be thrown, &c., &c. He was covered with perspiration, the arms and legs trembled ; he felt weak, had pain in his back, but without any tendency to syncope; the epigastrium was tender on pressure ; no nausea nor vomiting ; conjunctival ecchymosis of the right eye, blood-stained expectoration ; no râles in the lungs; heart-sounds well heard, the first a little harsh; urine clear, without traces of blood, heat and nitric acid revealed no albumen in it, nor did Bareswill's solution shew any sugar.

Treatment. A litre of water with white of egg. A dish of broth. Quinine wine four ounces, with an ounce of syrup of orange peel.

May 30th. Hallucinations persistent ; he is threatened; they seek to kill him; he is told so every minute, he sees armed men coming into his room.

Marked trembling of the hands.

Conjunctival ecchymosis in both eyes ; in the right the mucous membrane is raised over the lower *cul-de-sac* ; the face and more particularly the fronto-temporal regions are covered with ecchymotic dots of petechial character; in some places the spots are more extensive and upon the lips two are seen of the size of a sixpence. No nausea; the epigastrium is a little painful; there is tympanitis with resonance of the whole left hypochondrium ; some borborygmi without pain; nothing particular to be observed with regard to the liver; the stools hard and without traces of blood; the urine, examined by the microscope, shews neither blood globules nor fat cells nor casts, it contains neither albumen nor sugar.

Respiration normal, regular ; some scattered râles in the upper half of the right lung at the back; no dulness nor blowing breathing anywhere. The first sound of the heart is prolonged at the apex, nothing to be noted at the base nor in the carotids. Pulse regular, small, 80. The blood obtained by a prick made at the seat of an ecchymosis appears to be normal on microscopical examination.

Treatment. Quinine wine. Water with white of egg.

May 31st. The delirium continued the same during the day; general condition the same ; fresh ecchymoses on the sides of the neck.

During the night he is very restless. He gets out of bed and refuses to go back ; they want to smother him, he sees them making all preparations; forbids anybody to go near him.

June 1st. He is calmer, but the delirious ideas remain. He complains of no suffering. The pulse is small and regular, the first sound of the heart still appears to be prolonged at the apex.

The ecchymoses on the forehead are a little paler, the conjunctival ecchymoses are of a shining red colour. The expectoration is no longer stained with blood; the breathing is easy and natural. The epigastrium is scarcely at all tender. There is constipation.

Treatment. White of egg and water; quinine wine; an emollient enema.

June 3rd. Still hallucinations; he hears it said that they are going to fire at him with their guns; they will not let him be quiet, they are always reproaching him.

The ecchymoses on the forehead continue to diminish; those of the conjunctiva are not so red, no fresh ones appear. Treatment the same.

June 5th. He is calmer in the daytime; the hallucinations are less frequent, but the delirium persists. The general condition is good; the ecchymoses tend to disappear. The same treatment.

June 15th. He gets up in the day-time, is quiet, but retains the ideas of persecution.

The ecchymoses are gradually disappearing; there is a sub-icteric tint of all the skin. The hepatic region presents nothing particular.

July. The patient is quiet, and finds something to do, but his intelligence is clouded; notions of persecution remain, and the hallucinations still shew themselves at night.

The ecchymoses have disappeared.

September. H—— whose intellectual condition is somewhat lowered, still retains his confused ideas about being persecuted; during the night he has hallucinations of hearing and sight of an unpleasant character.

His general health is good; the skin retains a sub-icteric tint. He experiences no notable change until towards the month of May, 1869; at this time H—— sleeps better, is less sad, finds some work to do more willingly, but he has still delirious ideas which one discovers readily as soon as the conversation turns on subjects which had formerly occupied his attention the most. One notices, moreover, a certain indolence and slowness in his conceptions.

The patient went out improved but not cured on July 29th, 1869.

On his arrival at home he did not discontinue his excesses in drinking. The hallucinations and delirium returned with fresh intensity; he cried, shouted, threatened death and fire, and entered the asylum for the eighth time on October 24th, 1869.

On admission he is restless, hears persons threatening him, sees policemen following him, and wishing to arrest him; they wish to get rid of him, &c.

In spite of his agitation one can, nevertheless, easily get his attention, and obtain some answers.

His hands are tremulous, he sleeps four hours during the night; the next day the hallucinations are less frequent. Headache; dizziness; cramps in the limbs. Trembling of the hands.

At the end of a month the hallucinations are almost completely gone, but his intellect is weakened and his memory less precise; confused ideas of persecution remain and revolve on the same subject.

Dizziness; numbness and cramps in the legs.

In the month of May, 1870, H—— escaped from the asylum and was brought back two days afterwards in a state of complete drunkenness which was soon followed by violent excitement.

For five days he screamed, gesticulated, was frightened, and either defended himself or was abusive; then, having become quiet he again presented weakening of intellect and confused notions of persecution, headaches, dizziness, and cramps in the arms and legs. In the month of August he was sent to a provincial asylum.

In this case one can trace the progressive march of alcoholic intoxication and its more and more profound action on the organism. In proportion as the relapses increase in number, we see persisting after the disappearance of the acute or sub-acute symptoms of the poisoning, a cerebral malaise and intellectual torpor, which a little later is replaced by obvious impairment of the faculties; after the eighth admission we find, in fact, a marked tendency to dementia with physical phenomena which makes us fear the existence of already serious lesions; lesions induced at once by the alcoholic and the phosphoric intoxication. H—— has thus arrived by degrees at chronic alcoholism.

The poisoning by phosphorus exhibited some

very noteworthy peculiarities in our patient; not to mention the favourable termination, which is quite an exceptional occurrence, we have seen that blood-stained expectoration occurred on the third day together with conjunctival and cutaneous ecchymoses, which, after having increased for three days, and foreboded a fatal termination, were slowly effaced, to disappear without other symptoms. In the hæmorrhagic form of poisoning by phosphorus, M. Tardieu* mentions, as liable to occur during the first few days, vomitings of pure blood, and liquid bloody stools; but the cutaneous ecchymoses and petechiæ do not generally occur until the end of three weeks or a month, at the time when the multiple hæmorrhages appear in all directions. The premature occurrence of the cutaneous hæmorrhages was undoubtedly favoured in our patient by the pre-existing anatomical condition of the capillaries, their changes being analogous to those occurring in poisoning by phosphorus; but, whilst admitting this hypothesis, we must confess that we are not quite satisfied with it, seeing that the fatty degeneration of chronic alcoholism affects all the capillaries of the system, and that in our patient the capillaries of the face and neck were more particularly the seat of the extravasation of blood.

CASE XXIII. *Uncle insane.—Convulsive symptoms in infancy.— Abuse of wine, brandy, and absinthe for twenty-five years.—In the last twenty years several attacks of alcoholic delirium with fits of epilepsy and vertigo.—Weakening of the intellect, hypochondriacal ideas; tendency to dementia.—Motor and sensory disturbances, more marked on the left side; hemi-anæsthesia of the general sensibility and of the senses.*

Louis G——, æt. 43 years, a carpenter, came into St. Anne for

* Tardieu. *Etude médico-légale et clinique sur l'empoisonnement.* Paris, 1867, p. 436.

the fourth time on February 17th, 1872, after having been already treated several times in ordinary hospitals for alcoholic symptoms. His uncle, on the father's side, was insane, he died at the Bicêtre after a residence there of fifteen years. His father, who was a sober man, and enjoyed good health, died by an accident in 1848. His mother is healthy and still lives in a home for the aged.

G—— was fairly intelligent when a child, went to school and learnt with ease; from his twelfth to his sixteenth year he suffered from headaches, dizziness, vertigo, and even attacks of loss of consciousness, symptoms of such a character as led to the application of leeches to the arms. In his 17th year, having worked one year with his father, he began travelling for work; from this time his first excesses in drinking are dated. In 1848 he lost his father and found himself more at liberty to indulge his taste for spirituous liquors. In 1851, when twenty-two years old, having been exempted from military service as a widow's son he enlisted and joined the musical band and was sent to Africa. To wine and brandy he now added absinthe, of which he occasionally drank considerable quantities. He became gloomy and restless, the least opposition irritated him, he slept badly, dreams and nightmare disturbed and wearied him. He suffered from vertigo and several times he fell down, suddenly losing consciousness, struggling, biting his tongue and sometimes urinating into his trousers. *"Several of his comrades had attacks like his,* he said, *and they would often drink, three or four of them together, as much as a bottle of absinthe in a day."* In 1854 he contracted syphilis, which after an indurated chancre of the penis revealed itself by roseola, headache, osteoscopic pains, a papular eruption, onychia, exostoses and iritis of the left eye. In the same year he had an attack of intermittent fever. Weakened by his disease, he sought a remedy in absinthe and brandy, making free use of them. He suffered from frequent headaches, dazzling, dizziness, tremors, cramps in the arms and legs, vertigo, and fits occurring more especially when his excesses were more frequent. From 1858 to 1868, in which year he returned to France, he was admitted into hospitals several times, and each time, said he, he trembled in all his limbs, had visions, saw lions, cats, dogs, and armed Arabs; after some weeks' rest he rejoined his regiment to resume his former kind of life.

One fact well worthy of notice is the existence at this period of a very marked weakness of both legs lasting for a fortnight which was followed by numbness and cramps for several months.

On his return to Paris he worked at carpentry without abandoning his habits of drinking and his predilection for absinthe. In the course of the year 1869 he was twice admitted into the Necker Hospital with tremblings and hallucinations. At last on the 14th of January, 1870, he was brought to St. Anne. He had a stupid look, at times was in terror, saw animals, and bears which seemed to swell out and come near to devour him; he saw Arab women and children armed with poinards who wished to murder him; he saw rats and cats. On questioning him one could arrest his attention, but his answers were confused and his mental faculties appeared to be weakened.

Headache, dizziness, considerable trembling of the hands, muscular shocks in the arms and legs, and sometimes in the face. No heat of skin; pulse 76, regular; tongue coated; pains at the epigastrium.

Treatment.   A bottle of seidlitz water.

Two bottles of citric lemonade.

The night was disturbed by hallucinations, but towards the next day he was calm.

Frontal headache, dazzlings and vertigo; numbness of the limbs, cramps in the calves; slight cutaneous anæsthesia of the legs.

G—— was sent to an asylum in the country, where he remained until April 14th, 1870.

On his return to Paris, fresh excesses were followed by hallucinations, delirium, and epileptic attacks.

He was brought back to St. Anne on May 10th, 1870.

He had the same aspect as in the month of January; was stupid, answered questions put to him very confusedly, and at times was excited and frightened by his hallucinations. At the end of three days he was more tranquil, understood his condition better and related how after having drunk he wandered about at random and was found the next day lying in a dung cart, without knowing how he had got there.

He frequently suffered from vertigo, and since he went out the last time he had had two fits with loss of consciousness, convulsions, biting of the tongue and involuntary evacuations. G—— was sent to another asylum, where he remained nearly two months, after which he went out and enlisted during the continuance of the war, serving in the second Lancers in the army of the Loire.

This period apparently marked an arrest of this long poisoning, whether from a feeling of duty or from want of opportunity

G —— made trial of a comparative sobriety. When the peace was signed he left his regiment and returned to Paris ; his sad history began again, he again gave way to drinking, was seized with delirium and was brought back to St. Anne on July 26th, 1871.

He was sad and looked depressed, he occasionally heard insulting speeches, the night before he saw a man cut off three women's heads, he saw lights, birds, and animals which seemed to grow larger.

Heaviness of the head, dizziness, numbness of the limbs, weakness of the legs; after having sat down a short time he finds a difficulty in rising and feels great weakness in his knees; cramps in the calves; a pin stuck into his skin provokes but little pain; the temperature is found to be normal over the whole surface of the body.

G —— worked some hours in the daytime in the carpenter's shop; he ate with appetite, slept well, but his faculties were weakened. At the beginning of February, 1872, he left the asylum, but was brought back again a fortnight later.

He had made good use of his few days of freedom to compensate for the long abstinence from alcoholic drinks enforced upon him during his late confinement. On the very day he left he drank four glasses of absinthe ; the next day he took double the quantity, and in this way reached a daily dose of from fifteen to twenty glasses. The day before re-admission he had had an epileptic attack and was found hanging by his legs from the banisters of the staircase.

This time, again, hallucinations were present, he saw dogs running after him and biting him ; he saw grinning figures which seemed to enlarge and diminish ; and scaffolding which got loose and fell around him.

Trembling of the hands; pains in the elbows and knees; cramps in the calves, tingling in the plantar surface of the feet.

After some days the symptoms brought on by his late excesses disappeared, and G —— manifested physical and intellectual disturbances which denoted lesions already far advanced.

His faculties were weakened ; his memory less precise ; in his recital of the numerous symptoms he had experienced he gave erroneous dates, contradicted himself, told the same things several times over; he was apathetic, indifferent, he willingly granted that his conduct was blameable, but was far from rightly appreciating it. He was hypochondriacal, troubled beyond measure about insignificant ailments ; he slept badly, often had nightmare, frights and unpleasant dreams.

Heaviness of head, with a feeling of pressure on the forehead, beating in the temples, humming in the ears ; numbness of the limbs ; isolated and irreguiar shocks in some of the groups of muscles in the arms and especially in the legs.

Sometimes he had very painful cramp ; tingling of the soles of the feet, and on several occasions numbness and tingling in the left side.   The left arm was a little weaker than the right ; the sensibility, which was blunted over the whole surface of the body, was more obtuse on the left side, the pricks of a pin, pinching and tickling were felt less than on the right side.   From time to time he suffered from vertigo.   Still he had had no attacks of epilepsy since the time of his first admission.

With the rest, the regular mode of living in the asylum, and with the discontinuance of his excesses, G——'s health was sufficiently restored by the end of six months to enable him to do some carpenter's work, though not with the activity and cleverness of former times.   His appetite was pretty good ; his sleep disturbed from time to time by dreams and nightmare ; he had headache, numbness, and cramps in the limbs.

At the beginning of 1873, as the improvement persisted, the patient, to try him once more, was allowed to leave the asylum.

Once again his own master, G—— resumed his wild career, was dismissed from the workshops and spent the money he had just earned in drinking.

He was re-admitted into St. Anne May 9th 1873.

On admission he is dull and stupid, his faculties are notably impaired ; he is assailed by hallucinations at night; he sees soldiers, men armed with sticks, human forms, bear's heads, devils, which enlarge and grow less and change colour ; he hears the report of musketry.

Trembling of the hands ; numbness : cramps.

He appears to have had an epileptic attack, his tongue is deeply bitten, he has no recollection of what has passed.

The hemiplegia is well marked ; he exerts less force on the left side, the right hand reaches the 50th division of the dynamometer, the left only the 20th.

The sensibility is equally weakened on the left side, he scarcely feels the pricks of a pin, discerns badly the temperature of bodies, he feels however when one blows on his hand and arm.   On the right side the various modes of impression are exactly distinguished. He has throughout the left side a sensation of coldness which at times is disagreeable, yet by the aid of the thermometer a difference of only $\frac{1}{10}°$ C. less in the left is found between the two hands.

The right ear hears the beating of a watch at a distance of 20 inches; with the left ear, it is only heard at a distance of $6\frac{1}{2}$ inches.

Vision is less clear on the left side, there is also some defect in the recognition of colours on this side; but an old iritis aids the hemi-anæsthesia in disturbing vision on the left side, and it is difficult to determine the exact share due to one or other of these two pathological causes.

The left nostril detects the odour of neither orange flower water nor camphor; vinegar provokes a vague sort of sensation; on the right side these substances are easily recognised. The left side of the tongue distinguishes neither salt nor colocynth, which are rapidly and perfectly tasted on the right side. The patient had himself noticed this loss of taste on the left side while eating. There is anæsthesia of the buccal mucous membrane on the left side, titillation of the velum palati and uvula on this side provokes no nausea. The sensibility is equally diminished in the nostril, the external auditory meatus and the conjunctiva of the left side.

The patient has had a cough for some months; slight dulness and crepitation are found at the apex of the left lung.

The hemiplegia and hemi-anæsthesia improved by degrees, but giddiness, numbness and cramps persisted, especially on the left side.

The patient continued to cough, and spat up blood on several occasions. On November 4th, 1873, G—— escaped from the asylum; scarcely was he outside before he began to drink, and next day he was picked up in the street dead drunk.

He was brought back to St. Anne in a state of profound stupor which, however, passed off very quickly. He was ashamed of his conduct, and began upbraiding himself before one could speak to him. His sleep was disturbed by dreams and nightmare for two days. The left hemiplegia and hemi-anæsthesia seemed to be aggravated. The right hand reached the 40th degree of the dynamometer, the left only the 30th, the left leg was also enfeebled. The pricks of a pin were scarcely felt on the left side, but a breath as well as temperature were discerned, but not nearly so well as on the right side. The buccal and nasal mucous membrane, the conjunctiva, and the external auditory meatus were anœsthesic on the left side, on the right the patient felt the slightest touch.

The left nostril fails to recognize orange-flower water, essence of mint, musk, and camphor ; on the right side on the contrary these substances are immediately designated by name.

The left side of the tongue distinguishes tastes much more slowly and with less precision than the right.

With the left ear the ticking of a watch is heard at a distance of six inches ; with the right, at twenty-four inches.

The sight is also impaired on the left side.

Dizziness, dazzling, whistling in the ears ; numbness of the limbs, more marked on the left side.

Treatment. Vinous lemonade ; a draught with half a drachm of extract of cinchona and an Anderson's pill every other day.

G——, is not a simple alcoholic, he is a patient with a predisposition, whom we have ranked among chronic alcoholics because of the already grave symptoms which he presented, and which seemed to offer scarcely any chance of cure. His uncle was insane, and he himself had had convulsive symptoms when from 12 to 16 years of age ; of these we have not been able to obtain precise notes.

When 17 years old, G—— began to commit excesses, but it was chiefly from his 23rd year, when in Africa, that his abuse of brandy and absinthe became excessive. He soon began to experience attacks of alcoholic delirium, vertigo, and epilepsy, which became the more frequent, as his excesses multiplied. Attributing only a purely excitant action to the excess of drinking in this patient, we might consider the epileptic attacks to have been due to an hereditary influence. It is probable, indeed, that the latter exercised a certain influence in the development of the convulsive crises, but the excesses in absinthe were too closely followed by the attacks not to lay the greater part of the blame to them.

This case, moreover, is of special importance from the point of view of the progress of symptoms; we see the intelligence gradually decay and *pari passu* with this there appear more and more marked physical disturbances. The motor and sensory symptoms which appeared in the lower limbs on several occasions lead us to suspect a chronic irritative process in the cord (diffuse sclerosis), on the other hand the persistent weakness of the left side of the body with the hemi-anæsthesia of the general sensibility and of the special senses caused us to fear the existence of a lesion (probably hæmorrhage or ramollissement), in the right hemisphere. With these organic conditions the prognosis, we conceive, is much more serious.

CASE XXIV. *Old habits of drunkenness—Intellectual decay; headache, giddiness, cramps; two epileptiform attacks—In 1866, the first attack of alcoholic delirium, grafted on chronic alcoholism—From this time to 1872 admitted eight times into asylums after excesses, with sub-acute symptoms added to chronic alcoholism—Marked tendency to dementia.*

Francis A——, a tanner, aged 47 years, has been given to drinking for years, he has taken chiefly white wine and brandy; he has frequently suffered from headache, dizziness, and cramps in the calves; the intellectual faculties and memory are notably impaired. According to his wife's account he was suddenly seized with two convulsive attacks in 1860, but they have not recurred.

In 1866 he was for six weeks under treatment in a private asylum for an attack of alcoholic delirium.

On May 17th, 1867, he was admitted into St. Anne for the first time; he was excited, had incoherent notions, was angry at times, struck at and broke objects; he complained of headache, dizziness, pains in the stomach and vomiting.

For a fortnight he could hardly sleep and his rest was interrupted by hallucinations. He tried to drive away rats and cats which jumped upon his bed; he thought his room was full of smoke and mist; was constantly complaining of seeing lights, flames and sparks; the excitement was redoubled at night, it seemed to be connected with fears the cause of which could not be detected.

He was transferred to another house and spent some weeks there. He went out, resumed his excesses and was re-admitted into St. Anne, Sept. 9th, 1867. His memory, which was quite exact about former events, was often found to be at fault with reference to recent events.

What seemed to be predominant was a vague feeling of fear and sadness; the sight of a child or of a dog inspired the thought that the one was about to throw a stone at him, the other, to bite him. If he heard a clock strike he subsequently dreamt about his funeral; his life had become a burden to him, he wanted energy and impulse; ideas of suicide would frequently assail him but the force necessary to put them into execution failed him.

His fears increased at night, and his sleep was disturbed, but the remembrance of the indistinct hallucinations did not persist in his waking hours. He had marked trembling of the hands, the lips, and the tongue.

He went out better after several weeks' treatment in a country asylum. He returned to St. Anne for the third time at the beginning of June, 1868.

His nights were filled with terrifying hallucinations: "they wished to injure him, they wanted to kill him." Under the influence of such ideas he had fallen on his knees in the middle of the court of a slaughter house in which he worked, saying he was condemned to death, and begging for pardon.

The delirium was less active in the daytime than at night, but unpleasant hallucinations disturbed him; he saw strange forms and death's-heads.

We observed dizziness, humming in the ears; cramps in the hands and legs; itching over various parts of the skin.

Considerable trembling of the hands.

His sight was weak; the lens of the right eye was beginning to be opaque, whilst that of the left was normal. Atrophy of the choroid around the papilla on both sides.

Treatment. Aloes; a bath; quinine wine.

June 5th. The hallucinations have been very numerous in the night, he saw and heard people threatening him.

June 9th. To spend his time he tries to read; his eyes follow the letters but he cannot retain anything.

June 19th. His sleep has been much disturbed; words were uttered about him which he did not comprehend; he is surrounded by death's-heads. The hands and tongue still tremble much.

June 29th. The hallucinations have ceased, there remain pain, and at times, cramps in the limbs. With the dynamometer, the right hand reaches the 39th degree, the left the 42nd; in his ordinary work he uses the left hand.

He was sent away to an asylum, where he spent some time, subsequently he was re-admitted for the fourth time on May 27th, 1869.

He was in a state of slight excitement with some confused hallucinations, his intellectual faculties and memory were much impaired. Dissatisfied with himself and with others, he spoke of putting an end to his life and of killing his wife, but without expressing this wish by any overt act.

He was admitted for the fifth time on August 4th, 1871.

His intelligence appeared to be decaying more and more, he was incapable of work and of planning anything, his ideas were confused, he shewed very indistinctly whatever he experienced physically or intellectually. Having left the asylum to which he had been transferred, on June 11th, he was shut up in St. Anne for the sixth time Sept. 27th, 1871.

At this time one noticed the existence of certain painful hallucinations; in the night he was surrounded by spectres, apparitions and frightful pictures, he suffered from cramps and itching all over the body.

Having left fifteen days previously, he was brought back for the seventh time on June 6th, 1872. He had gone into the country, but had immediately given way to drinking white wine and liqueurs. Seized with despair at the thought of his physical sufferings, and at the remembrance of the painful hallucinations which so frequently disturbed his nights, he was seen to weep and no answer could be obtained from him, but that he was a wretched fellow unfit to live and unworthy that henceforth anybody should trouble to take any interest in him. His physical sufferings had increased and appeared to be very severe, at times he had acute pain in the pit of the stomach with very painful cramps, coming on in paroxysms; vomiting, loss of appetite. He was very retired, spoke but little and contented himself with groaning when asked to explain what he felt. His intelligence failed still more, and when in his 52nd year R—— gave signs of advanced dementia.

R—— is an alcoholic in whom the disease has become chronic; his delirium and hallucinations resemble those of the most ordinary cases.

At first with some weeks of confinement and with very simple treatment improvement was rapid. The various doctors who treated him agreed in restoring him to liberty as they saw the delirium pass away. Fresh excesses brought him back time after time until at his last admission he presented himself in a state of advanced dementia with gastric mischief, and with a disturbance of motor power which indicated organic lesion, the fatal result of long years of excess in alcoholic drinks.

ARTICLE IV.

Cases of chronic alcoholism terminating in general paralysis.

Authors have not failed to assign an important place to alcoholism in the etiology of general paralysis, and what we have said of the prolonged action of alcoholic drinks on the animal economy explains the steps by which in certain cases alcoholism passes into paralysis. This passage does not occur all at once, there is an intermediate or transition period which is sometimes very long, and difficult to estimate precisely, which compels us to suspend our diagnosis; Case xxviii is a good example of this.

In these cases, as we have seen, there is superadded to the fatty degeneration of the organs and athcroma of the vessels, a diffuse interstitial sclerosis; nor does this chronic inflammation of the envelope and interstitial tissue exclusively affect the brain; other organs, for example the spinal cord, liver, kidneys and heart may likewise become its seat.

The four following cases will suffice, we think, to give an idea of cases of chronic alcoholism terminating in general paralysis. In Cases xxv and xxvi, we find occurring simultaneously cirrhosis of the liver and diffuse interstitial myelitis with more marked localization in the posterior columns.

In Cases xxvii and xxviii, after several attacks of simple alcoholic delirium, the weakening of the faculties, and some disturbances of a new character in the motor functions, lead us to recognize the fact that the disease is entering upon a new phase, and this transition period leads finally to general paralysis. The autopsy shews, in addition to fatty degeneration of the viscera, many traces of chronic inflammation.

CASE XXV. *Alcoholic excesses;—Cerebral disturbances. Spinal symptoms of old date. General paralysis with ataxic disorder of the lower limbs.—Autopsy.—Lesions of general paralysis; diffuse interstitial myelitis with localization in the posterior columns.—Visceral lesions of chronic alcoholism. Fatty degeneration of the heart, arteries and kidneys; cirrhosis of the liver.*

Jules V —— aged 50, formerly a soldier, came into St. Anne on June 8th, 1868.

For a long time he had been given to drinking. Ten years ago he had an attack of mania, with painful hallucinations and suicidal ideas; from that time his character changed, and he had fits of passion and sometimes of violence; later and at a time which cannot be exactly determined, he had headache, attacks of giddiness, cramps in the calves, and then a sense of weight in the legs and difficulty in walking.

In June, 1868. Weakening of the intellectual faculties, ambitious delirium of exaggerated character, inconsiderate purchases. Hesitation in speech, trembling of the tongue and lips; tottering, jerky walk, which is more difficult with the eyes shut; the legs are widely separated and the foot thrown out in an irregular manner, sometimes overstepping the mark, at others falling short of it.   Lancinating and wandering pains in the lower limbs.

On Aug. 24th, 1868, epileptiform attacks followed by coma and death.

Autopsy.   Cerebral meninges slightly œdematous with some opaline spots in places; sanguineous infiltration forming a layer 5 to 6 centimetres in width in the meshes of the pia mater over the right hemisphere.   Adhesions, especially at the base and towards the temporo-sphenoidal lobes; meninges of the cerebellum thin and non-adherent; granulations in the fourth ventricle.   The large arteries are atheromatous, the basilar much dilated at the junction of the two vertebrals; the meninges of the cord slightly œdematous.   The posterior columns present a grey gelatinous tint throughout their whole extent, but it is much more marked at the lower part of the dorsal region, where it appears also in the posterior part of the lateral columns.   Heart large, fat, of brownish-yellow colour.

Aorta atheromatous, dilated in the transverse portion of the arch. Liver granular, cirrhotic.   Kidneys voluminous, slightly yellow in their cortical portion.

Microscopic examination of the altered parts of the cord shewed, in the fresh condition, fine isolated granules, some of which were united into irregular granular bodies, scattered over the section and accumulated along the vessels; on staining these preparations slightly with carmine, a considerable number of nuclei were observed, mostly elongated, some in the midst of a finely striated substance.

After hardening in chromic acid, one could better appreciate, in thin sections, the sclerosis of the posterior columns, which extended also into the lateral columns in the dorsal region.   Just below the level of the medulla only a triangular tongue of sclerosis situate on the inner aspect of the posterior columns was found, which terminated in a point at the anterior two thirds of the posterior median fissure.   There appeared, moreover, around the cord in these parts, an annular thickening of the connective tissue, with swelling opposite the extremity of the posterior columns.

In the dorsal region, the ring of sclerosis subjacent to the pia mater was much less thick.

The commencement of the symptoms in Jules V—— was of very old date, and the excesses in alcohol appear to have played the principal part in the production of the disease. We would remark the co-existence of sclerosis of the cord and cirrhosis of the liver. Such cases, of which several are now on record, are not the result of a simple coincidence, but are a proof of the existence of the same pathological process in different organs under the influence of similar causes.

CASE XXVI. *Habitual alcoholic excesses—Numbness and weakness in the limbs, followed by symptoms of general paralysis—Autopsy—Diffuse sclerosis of the cord, with predominant affection of the posterior columns; diffuse interstitial encephalitis; cirrhosis of the liver; atheromatous plates on the vessels.*

Louis B—— aged 45 years, a teacher of billiards, had for a long time indulged in excessive drinking; for three years past he had suffered from numbness and pains in the legs which recurred at irregular intervals; the arms were also sometimes numb. For some months, walking has become fatiguing and difficult.

On entering the examining office on April 19th, 1869, Louis B—— had weakness of the intellectual faculties together with ambitious delusions, hesitation in speech, trembling of the limbs, and inequality of the pupils; the legs, which were anasarcous up to the knees, were feeble; the walk tottering without presenting the characters of ataxy; the sensibility remained normal. The urine contained neither albumen nor sugar. The symptoms of general paralysis became progressively more marked, and paraplegia with œdema of the lower limbs increased more and more.

At the end of November, an attack of pneumonia occurred which carried off the patient on Dec. 4th.

Autopsy. Atheromatous plates on the arteries at the base of the brain; meninges thick, opalescent, adherent, ependyma of the floor of the fourth ventricle roughened, some granulations also on the surface of the lateral ventricles; successive sections of the hemispheres show nothing abnormal in the white centres; the meninges of the cord are pale and infiltrated with serous fluid; sections made in the different regions shew a diffused greyish tint, most marked in the posterior columns. Both lungs engorged, the right dense, friable, and hepatized at the base.

Heart soft, loaded with fat; mitral and aortic valves present at their margins small red vegetations, which are somewhat firm and of polypoid form.

The aorta is yellowish and has atheromatous plates here and there.

Cirrhosis of the liver; the surface of the organ is mammillated, the capsule thickened and the tissue very granular.

Kidneys fatty and of yellowish colour.

Fresh preparations of the spinal cord exhibit granules and granular bodies, scattered throughout the preparations and on the walls of the capillaries; their number is very considerable in preparations taken from the inner part of the posterior columns. In sections made after hardening in chromic acid, a diffuse sclerosis of the anterior and lateral columns is seen, the change, however, is much more marked in the inner part of the posterior columns.

In the brain were the lesions usually found in general paralysis.

Alcoholism played an important part in this case, and as a source of irritation it no doubt contributed to the development of the general paralysis, but in addition to the lesions in the cerebro-spinal axis, we found in the liver, kidneys, and vascular system, the more special changes of chronic alcoholism.

We might also in this case have met with softening of the brain from embolism, due to the condition of the mitral and aortic valves in the patient, but that would merely have been an extra lesion, which by giving rise to symptoms of hemiplegia, might have complicated and partly hidden the features of the principal disease.

It will be useful for us to bear in mind these pathological combinations, in order to analyse and estimate at their true value certain facts which would otherwise appear very contradictory.

CASE XXVII. *Habits of drunkenness; first attack of alcoholic delirium on Sept. 21st, 1865; cure. Continued abuse of alcohol; digestive disorder: change of character; second attack of alcoholic delirium on Sept. 22nd, 1869.—Improvement incomplete, with per-*

*sistence of weakened intelligence; fresh excesses; third attack of tran-
sitory alcoholic delirium, leaving behind it obvious symptoms of general
paralysis. Autopsy—Lesions of general paralysis, and fatty degeneration of the liver and heart.*

Alexander V—— 43 years of age, was brought to the Bicêtre
on Sept. 21st, 1865, with the following medical certificate.
"Chronic alcoholism; acute attack of *delirium tremens* dating
from a few days back." V—— retained for several days some
hallucinations with ideas of persecution; slight embarrassment of
speech and some trembling of the lips. He improved pretty
quickly, and went out of the asylum on Oct. 7th, "in a good
state of health, both physical and moral" according to the certificate of the physician in charge.

After his discharge from the asylum, he resumed his work
and did it pretty well; but he soon began to drink to excess
again, addicting himself especially to wine and brandy; as a
result of these excesses, his health became obviously deteriorated,
and digestion difficult; there was epigastric pain, and the patient brought up strings of mucus in the morning; he suffered
from heaviness of the head, numbness in the arms and hands, and
weakness in the legs.

At the early part of 1869, he became odd, quarrelsome, irritable, and sometimes violent and dangerous to his neighbours; he
did not do his work so well, and his intelligence became lower.

On Sept. 22nd, 1869, he was brought back again to the Asylum in an attack of alcoholic delirium; he was a prey to hallucinations; he thought himself insulted and threatened; he saw
insects dancing about him, rats, cats and animals of all sorts,
flames and flashes of light. His whole body trembled; he looked
about him everywhere and had not a moment's rest. His face
was puffy and presented a slight icteric tint. The urine contained
neither albumen nor sugar.

On the 25th, it was noted. "The hallucinations are somewhat
changed in character; ideas of persecution predominating; V——
appears less terrified, but is still restless and distrustful; he hears
voices which he knows and thinks they are plotting against
him; his wife wants to destroy him; she is deceiving him; he
calls for her and demands her arrest. Intelligence notably
weakened. The quivering of the whole body has ceased, but there
remains considerable trembling of the hands."

Three days later, the hallucinations were occasional and in-

distinct, and allowed the patient more rest; and having become
calm, he was sent to a farm asylum where he remained till the end
of October.

He had hardly regained his liberty before he returned to his
drunken habits; he was no longer fit for labour, he did not know
how to do his work, and his faculties were markedly weakened.

He was brought back to the asylum on Dec. 16th, 1869, after a
fresh attack of alcoholic delirium.

On admission, V—— is terrified by numerous hallucinations;
he sees women, and armed men pursuing him to kill him; he is
insulted and threatened.   The speech is jerky, the lips, and
especially also the hands are trembling, and the left pupil larger
than the right.

Two days later. The hallucinations have almost entirely disappeared, the patient is calm and answers questions; it is now
easy to convince one's self of the weakening of his intellect, the
marked loss of memory, and the existence of hypochondriacal
ideas.   In addition to hesitation in speech, trembling of the
tongue and lips and inequality of the pupils are observed.   These
symptoms of general paralysis persisted till the time of death,
which occurred on Jan. 9th, from an attack of pneumonia.

The *autopsy* made on the 10th, shewed most characteristic
lesions; thickening and numerous adhesions of the meninges,
softening of the cortical substance, and thickening of the ependyma
which had a roughened appearance.   The heart was soft and
loaded with fat, the liver yellowish in places, the kidneys not
appreciably altered.

The direct action of alcoholic liquors is here displayed in the most striking manner.   A first attack
of alcoholic delirium is followed by cure, all the
symptoms developing as in ordinary alcoholism
without giving rise to any particular phenomena;
but the excesses are continued and the health is impaired and intelligence lowered.   A second attack
is brought on, and this time it is followed by incomplete recovery with a persistent weakening of
the faculties, which prevents regular work and shews
that a diseased condition of the brain has already been
set up.   In fact at the third recurrence of the affection,

the attack of alcoholic mania becomes but an acces-
sory element; the hallucinations rapidly disappear,
the trembling of the hands which was very marked
on the first day also diminishes, and, in a word, the
alcoholic symptoms decline one by one, leaving un-
masked the evident signs of general paralysis.

We will give one more case of one of our patients
whose clinical history has already been related elsewhere, and which we complete here, as the patient
has since that time succumbed to the progress of
general paralysis.*

CASE XXVIII. *Habits of Drunkenness since* 1856; *first admission to St. Anne on June* 14*th,* 1869 ; *alcoholic delirium ; cure.—
Second admission on April* 19*th,* 1871; *alcoholic delirium; slight
impairment of intelligence; ideas of self-satisfaction.—Third entrance on Sept.* 11*th,* 1871 ; *alcoholic delirium; weakened intellect;
apoplectiform and epileptiform attacks,* (intermediate period.)—
Fourth entrance on Nov. 30*th,* 1872 ; *intellectual weakness; ambitious delusions; hesitation in speech, inequality of the pupils* (obvious
general paralysis)*.—Autopsy.—Fatty degeneration of the liver, kidneys and heart; atheroma of the vessels; pericarditis; perinephritis;
cerebral pachymeningitis; chronic diffuse interstitial encephalitis;
interstitial neuritis of the* 2*nd pair of cranial nerves.*

Adolphe B——, a wine merchant, aged 42 years, came into
St. Anne Asylum for the first time on June 14th, 1869, with a
certificate from M. Lasegue as follows; "subacute alcoholic delirium of six days duration ; nocturnal excitement; an epileptic
attack."

On his entrance, the patient presents the characteristic symptoms
of acute alcoholism.   He sees animals of all sorts, cats, rats, serpents, and people who want to assassinate him.   Red flames pass
before his eyes; he thinks he sees a thick smoke coming out of
his mouth and spreading through the room.

* Gambus, *De l'alcoolisme chronique terminé par paralysie générale,* (Thèse de Paris, 1873, p. 32).   Lolliot, *De l'alcoolisme chronique comme cause de la paralysie générale,* Gazette des hôpitaux,
Sept. 6th and 13th, 1873.   (Memoir followed by a report by M.
Motet to the Medical Society of Paris.)

He complains of a severe headache, and presents a considerable trembling of the hands, tongue and lips, with a sub-icteric tint of skin.   The sight, moreover, is somewhat markedly weakened, but ophthalmoscopic examination shews nothing abnormal.

His wife gives the following information :   Her husband, she says, has long been given to excess in drinking ; he drinks chiefly bitters, vermouth and wine, and has done so since the year 1856. His sleep is rather frequently disturbed, and he has at times some expectoration in the morning ; and for some time, he has trembled a little and been subject to abundant sweating.—His doctor, ascribing the disease to excessive drinking, had tried to persuade him not to drink ; but it was of no use.   He did not leave off drinking until he felt ill, and he took to his excesses again as soon as ever he got a little better.

At the commencement of last May (1869), he had great agitation, and became subject to excitement, especially at night.   Some days later, three weeks before his admission to the ayslum, he was seized with a convulsive attack accompanied by loss of consciousness, after which his speech remained embarrassed for three hours.   The other day, whilst reading the paper, he noticed that he read five lines at a time.   And lastly, under the influence of hallucination and ideas of persecution, he frequently indulged in threats against his wife.

His father died at the age of 74, hemiplegic.   His mother, who died at the age of 71, was living with a soldier 47 years old as his wife at the time of her death. Three of his children died early of convulsions.   His fourth child, a little girl three years old, is well.   Three years ago, his wife, who is about 40, had a miscarriage of twins at about the fourth month.

On June 15th, that is to say, two days after his admission, he appeared to be rather excited ; and walked to and fro in an animated manner.   He imagined that he was spoken to, that he was obliged to walk, and that people were taking hold of him, insulting him, and keeping him from sleeping.   He had then marked trembling of the tongue and hands.

He was sent to the asylum at Vaucluse on June 17th, where he improved rapidly, and whence he departed, at the end of July of the same year.   On his return home, he resumed his work, and did not shew any change in character, except that he became very economical, almost a miser.   He did not stammer, nor appear to have the slightest hesitation in his speech,

This man came into St. Anne for the second time, on April 19th, 1871 ; nearly two years after his first discharge.

On admission he has again alcoholic delirium with hallucinations.   Free-masons, he says, are pursuing him, men armed with knives are running after him ; he sees birds, cats, rats, trees, &c.   He has, moreover, slight trembling of the hands.

His wife furnishes the following information ; he has given himself up to habitual excess; but, eight days ago, having left his wife, he set himself to drink heavily with a young girl eighteen years old whose company he was keeping at her house.   Then the delirium broke out, but he had been getting drunk every day for quite a month, chiefly on brandy.   After he went out of St. Anne, he appeared well, and had set himself to work again ; but for the last month there has been complete disorder, he carried off money, furniture, and everything from his house and sold it all.

August 21st.   Invisible free-masons threaten to kill him.   His bearing is but little suited to his age.   He has an imperfect consciousness of all the follies he has committed, and smilingly and complacently tells how he could easily begin them again.   There are evidently ideas of self-satisfaction in him.   He is about to go and buy some horses, some varnished boots, and a white felt hat ; he will eat oysters and live on his rents.   He has at least two-thousand francs of his own.   His mother, he adds, lives with a lover who has thirty thousand francs; "*he'll get all that for himself.*"

The acute symptoms of alcoholism and the notions of satisfaction disappear little by little, and he leaves the asylum towards the end of September, retaining a slight weakening of the intellect, apathy and indifference.   Nevertheless, he is calm, and is therefore given up to his wife, on trial.

Twelve days after his discharge, he comes back again for the third time on Oct. 11th, 1871.   There is now discovered a weakening of the faculties, diminution of memory, slight excitement with hallucinations ; he sees phantoms, herds of ferocious beasts with horns, and frightful sights.   The trembling of the hands is very marked.

His wife relates, how that during the twelve days that he was at liberty, he did not know what he was about.   On the eighth day, whilst at table, he became very pale all of a sudden, and could not utter a word.   In the course of the night, at repeated intervals, he had slight spasms of the face with grimaces and distortion of the mouth (epileptiform attack). On the following day he was seized with an attack of delirium. His eyes were haggard, he insulted those about him, and was absolutely determined to go out. He had also frequent hallucinations. His wife affirms that since he left the asylum he had not had any fresh excesses.

On October 13th, the day after his admission, he goes to and fro in the room as if he did not quite know what he was about. He picks at things, picks up the most various objects, puts some grass in his button hole; sees ghosts and birds. Generally speaking, he remains indifferent to all that surrounds him, eats gluttonously, and only complains of a little pain in his head. In other respects he appears to have no consciousness of his condition.

On the 17th he is seized in the middle of the night with great terror, and utters piercing cries. He sees white and black bears, ghosts, and all sorts of things.

He remained in much the same condition till the month of March. In April, having become much more calm, he was restored to his wife. At this period he had still a very marked weakening of the intelligence, with considerable diminution of memory.

He returned to St. Anne for the fourth time on Oct. 30th, 1872, eighth months after his last discharge.

He still presents great weakening of the mental faculties, diminished memory, and incoherent ambitious ideas. He is rich and possesses sixty-thousand francs; he has bought a chateau, and will set up a porcelain building there ; he was the chief workman; his parents are very rich. He presents also considerable trembling of the hands, and slight hesitation in speech.

His daughter relates that for five or six months after his discharge, he remained apathetic and indifferent, not concerning himself at all about his house, and that he afterwards became wandering. Of late he has had ideas of being rich. He said he was Emperor or King, and thought himself very rich. His wife confirms this information, and adds that he was incapable of work; he did nothing and complained of nothing; they treated him like a child. About a week ago he became excited and had slight delirium. He has had ambitious ideas; and he escaped from the house clothed, and was found in the street half-naked. For a long time he was not able to drink any more, never having any money, and being under constant supervision. He has not had any fresh epileptic attacks since leaving. His speech has gradually become hesitating.

Nov. 1st. He says he will go to Versailles to see the President of the Republic. He appears indifferent and incoherent, answers with difficulty and presents marked trembling of the hands.

During the following days, his ambitious ideas become more and more strikingly marked. He is son of Napoleon Bonaparte, he is Emperor, master of France, and will soon go to Bordeaux, his capital, where he will appoint emperors, &c. He is rich, possessing five hundred mansions, golden covers, a four horse carriage, &c. He enlarges complacently on his physical strength. "He is as good as the *Pont-Neuf;* he is strong and can carry five-hundred pounds with his arm stretched out; a hare could not catch him at a race; he has twelve children. He is very skilful, never misses a stroke at billiards, sings and dances enchantingly, and is cock of the walk at the opera".

Later, in addition to his former ambitious ideas of his strength, the mansions he posesses, his position as Emperor, &c., there are added some new ones. He is going to beat the Prussians, he will march at the head of the army, he will have two-hundred cannon, will appoint his minister of finances and save France.

The hesitation in speech is very pronounced; trembling of the hands considerable, but more marked in the right. The general sensibility is much blunted; one can stick pins into his skin without his withdrawing his hand. He smells odours pretty readily, distinguishes salt and sugar by taste, but does not recognize the bitterness of colocynth till the end of some moments. He appears equally strong on both sides, and jumps equally well with both legs. The right pupil is slightly more contracted than the left.

The memory is very obtuse, and he does not appear to remember the occupation of his early years. His nights are as excited as before, he calls out and sings, thinks he is at the opera, and on waking, has not the slightest recollection of it. From time to time he complains of pains in his head.

Towards the 10th of November, hypochondriacal notions become mingled with his ambitious ideas. He has only nerves, no bones, and that troubles him. Nevertheless, he thinks himself very strong, and on the whole, appears rather satisfied with his condition than otherwise.

On the morning of the 6th of January, the attendant noticed some spasmodic movements in the face, arms and legs, resembling those excited by electric shocks (epileptiform attack); and at the time of the visit the patient complained of headache, dizziness, cramp, and a sense of fatigue. On the following days, the countenance was expressive of pain, and the patient was gloomy, and complained of cramps and numbness in the right arm.

His ideas are confused, he is conscious of suffering but cannot understand what it is he feels. His walk is staggering, like that of a drunken man, he has almost fallen several times, and is now afraid of walking, for fear of falling. He complains also of headache, numbness, and painful sensations.

Notwithstanding the persistence of the ambitious ideas, he is at this time in a period of depression, which lasts for some days, and improves a little afterwards.

March 12th. He is calm, answers questions addressed to him willingly, and presents a notable weakening of the faculties with loss of memory. He does not know the day, the month, or the year. He does not know how long he has been at the Asylum, and he appears apathetic and indifferent. Yet he is happy and is satisfied with his confinement. He tells how he is about to start for Bordeaux, where he will be King or Emperor. He will have a hundred thousand francs, a million; he will take his doctor away with him to reward him for his kind care, and when he is asked what his father and mother are doing, he answers that they are wine merchants, and that he himself is also a wine merchant.

He boasts of his physical and intellectual qualities; says he is strong and vigorous, and can play on several instruments, dances very well, and if one asks him he forthwith begins to skip about. He thinks himself very well dressed, and undresses himself to shew with great satisfaction his drawers, stockings, asylum uniform, &c.

His speech is hesitating, and as soon as he tries to speak, his lips are seized with tremor; the orbicularis muscle is the seat of slight contraction. The tongue when put out presents a slight tremulous movement at the edge and tip. When the hands are stretched out they tremble slightly. The muscular strength does not appear notably enfeebled; he grasps pretty firmly and equally with both hands. His walk, which just at first appears normal presents at times little jerks which are more marked and more easily observed when the patient walks slowly than when he goes quickly or runs.

Sensibility to touch is almost normal; he feels when his hand is breathed upon; but the sensibility to pain is much diminished, he does not feel pricks with a pin even when deep. He eats gluttonously, and mentions with satisfaction the good things his wife brings him from outside.

During the end of the month of March, a maudlin condition becomes super-added to his former state; the patient begins to cry without any cause, and if asked the cause of his sadness he answers that it is ennui, that he will die in the asylum, and that he does not want to stay any longer.

Ophthalmoscopic examination, made about this time, shows that in the left eye the papilla is pale, especially in the centre, with extremely narrow vessels; in the right eye, there is a physiological papilla, slightly pale in the inner part, but much less so than in the opposite eye; the vessels normal.

On the morning of April 1st, the patient is obtuse, and does not seem to understand questions. The rectal temperature is 38·2° C. (101° F.); pulse 66.

April 2nd. Agitation all night; incoherent cries; he gets up, and stiffens himself, and grinds his teeth. Pulse 72. He shews great obtusion of intellect; hardly understands anything, articulates with difficulty. The pupils are dilated, right more than left.

April 5th. He had an epileptiform attack at about 3 a.m., with deviation of the head to the right, convulsive movements in the arm and leg of the left side, and relaxation of the limbs on the right side.

April 7th. Another epileptiform attack during the morning with predominance of movements on the right side. Temperature 38° C. (100·4° F.) falling to 37·6° C. (99·8°) on the following day; when the patient has had no fresh attack and is less stupid.

April 15th. Considerable depression of the faculties, no consciousness of his situation. Considerable embarrassment of speech, complete loss of sensibility to pain; pronounced weakness of the right side. The patient stands less easily on the limb of this side, and his right hand he cannot use at all.

May 7th. He tries with all his might to change his clothes, to go to Bordeaux. Still some hesitation in the speech; left pupil largest. The index of the dynamometer is moved to 39° by the right hand, and to 45° by the left. The patient shews, moreover, automatic excitement and utters inarticulate cries.

From the first of May, iodide of potassium was given in doses of 1 gramme (15½ grains) a day.

During the same month, the two following attacks were observed. On the 14th, at 3 a.m. an epileptiform attack occurred. The patient uttered cries, lost consciousness, struggled, and according to the statement of the attendant, remained in a stupid condition after the fit. On the next day at the visit, his speech is embarrassed, he is confused, begins to smile, is obtuse, and cannot describe anything of what he has undergone. His muscular strength seems rather greater on the left than on the right side; he holds himself upright well, and walks without tottering.

May 20th. He has torn off his moustache, saying, "I am no longer emperor, and I must not have moustaches any longer". The left pupil is slightly more dilated.

June. Progressive weakening of the faculties, embarrassment of speech, he eats gluttonously, and takes his food with his fingers; is dirty in his habits.

June 9th. Diarrhœa with some traces of blood.—Emaciation.

The use of iodide of potassium is discontinued. The diarrhœa continues for some days following.

June 17th. Intellectual faculties notably weakened. The patient has no idea of his condition; does not answer questions; from time to time he pronounces words which indicate ambitious delusions, such as castle, emperor, millions. As to other symptoms, there is complete incoherence; marked embarrassment of speech; inequality of the pupils; and he besmears himself with fæces.

In the latter part of the month of June, nothing particular was noted. His condition remained almost exactly the same in degree. When one questioned the patient, or tried to attract his attention, he uttered some words: "yes, yes, 'tis so." And if we pressed him further one sometimes saw him undergo a slight fit, during which he shut his mouth, and held his breath, his face reddened, his eyes were injected and became motionless and prominent; then, after only a few seconds, all came to a sudden stop, and he began to cry.

During the beginning of July, the condition of dementia continued, and underwent slight increase.

July 5th. He whistles, blows, utters cries, sings in a spluttering fashion, claps his hands without any cause, is dirty, smears himself with his excrements. The right pupil is still most dilated.

July 7th. Complete indifference to everything and everybody. He no longer recognizes his wife, nor his sister, and he does not

speak to them when they come to see him. He answers different questions addressed to him only by such words as: "Yes, I
don't know; 'tis so." Except these answers one can get nothing
out of him. Sometimes he flies into a passion without cause, he
groans, bites his lips or cries, always without cause.

He does not speak now of Bordeaux, a word which before he
had constantly on his lips, and if we speak to him, and say, " B——
we are going to Bordeaux," he always answers: "yes, just so."
He tries to tear off his clothes with his teeth; and is with difficulty prevented from besmearing himself.

As regards his physical strength, it is still pretty well maintained. Thus, during the day, he is able to stay up, go down
stairs to the infirmary, and even walk for some time, but always
without knowing what he is about. He is obliged to be dressed,
otherwise he would put the wrong garment on first, or inside
out, &c.

The functions of vegetative life are fairly well maintained. The
patient eats with great voracity, swallows gluttonously, and takes
everything with his fingers.

He habitually allows his motions to pass, although they are
sufficiently solid, he smears himself with them, &c.; passes water
in bed and never in the utensil, although precautions are taken
to prevent it.

During the night, he generally sleeps but little, cries out, and
throws off the bed clothes. In the day time he often remains
seated, and usually sleeps after meals. It is extremely difficult
to explore the condition of the general sensibility of the patient,
his answers to different questions being always : "yes, yes, so it
is." But the sensibility to touch and to pain seem considerably
diminished, as may be judged by the absence of reflex movements.

From this time to the 20th of July, his condition remained
stationary. At the latter date there is a note. His intelligence
is much weakened and he remains in a state of entire indifference
to all around him. He is seated on a sofa, his head bowed down,
half drowsy, not replying to any question, and manifesting no
desire and no feeling. Sensibility to touch and pain can hardly
be said to exist now, except in the lower limbs, which the patient
withdraws when they are pricked. In the face and neck we can
pass a pin through a fold of skin from one place to another, without B——'s appearing to have any idea of it.

The condition of the different senses can hardly be estimated,
on account of the advanced degree of dementia. As regards taste

and smell, they appear considerably weakened, and one is obliged to employ substances of penetrating odour and pungent flavour to obtain any result. Thus, essence of mustard, when presented to the two nostrils in succession, causes at most only a slight movement of reaction, and this result is perhaps due to the sensation produced at the same time in the conjunctiva, the eyes filling with tears. On applying pimento to the tongue for some seconds, the patient responds at last only by a slight groaning.

The urine examined by heat, nitric acid, and Bareswill's fluid gave only a negative result as regards albumen or sugar.

August and September. The dementia continues to progress; the patient is excited, cries at times, and throws his arms about; he appears to have from time to time painful hallucinations, and pronounces the words brigand, brute, assassin, in the middle of the night and sometimes in the day; at other times, there are words which relate to his ambitious delirium; million, emperor, castle. Speech is embarrassed. One cannot get any sensible answers; he appears to be lost to all that goes on around him; his instincts and appetites are gone, he asks neither for meat nor drink, and swallows whatever happens to be on his plate; he is dirty and befouls himself with excrement. At the end of August, after an epileptiform attack, the right side remains weaker for four days.

At the end of December, digestion is disturbed, and diarrhœa supervenes; bed-sores occur over the sacrum. From Sept. 25th, to Oct. 4th, the pulse becomes small and frequent, and the temperature in the rectum rises progressively to $39°$ C. ($102.5°$ F.) $40°$ C. ($104°$ F.), and even $40.8°$ C. ($105.4°$ F.), with slight exacerbation in the evening. From Sept. 26th, the right side of the chest gives a dull note in its lower half, with bronchial breathing and fine subcrepitant râle. The strength declines, respiration is impeded, and death occurs on Oct. 4th.

During the course of 1873, the vital temperature was taken regularly night and morning for several months, and it shewed several times, a transient elevation, which each time it occurred was subsequent either to attacks of dizziness with pallor of the face and a slight apoplectiform seizure, or after slight epileptiform attacks; and after 24 to 48 hours had elapsed, the temperature was found again to be at its usual limit of $37.2°$ C. to $37.6°$ C. ($99°$ to $99.8°$ F.).

In this case, we took the tracing of the pulse with the sphygmograph several times, always with the same result; and we also reproduce here the tracing from the thesis of M. Gambus, which gives an exact idea of the state of the circulation of the patient.

Trace I.—Pulse of B——; CASE XXVIII; left arm; July 19th, 1873.

We have kept the letters of B——, written at different periods of his disease, so as to follow the progressive march of the dementia by the aid of these documents; and in the work of M. Gambus, the facsimiles of three of these letters may be seen : the last is a formless mixture of strokes, lines, and circles, with an enormous blot.

Trace II.—Pulse of a general paralytic without alcoholic antecedents.°

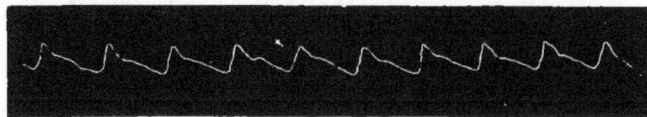

Autopsy, 26 hours after death. The walls of the cranium are thick, the diplöe reddened and injected. The dura mater is thickened and presents a false membrane 6 to 8 centimetres ($2\frac{1}{2}$ to 3 inches) in extent over the middle of the convex surface of the left hemisphere; this false membrane is easily detached by forceps, and when held spread out against the light is seen with the naked eye to contain vessels full of blood. A second false membrane is found over the right occipital fossa in the situation corresponding to the inferior surface of the right hemisphere of the cerebellum ; it tears into thin layers on attempting to remove it.

The arachnoid and pia mater are thickened, milky white in some places, and vividly injected in other parts, where they present

° In contrast with the sphygmographic tracing of B——, who became a general paralytic, as a result of the abuse of alcohol, we place the tracing of another patient affected with general paralysis without alcoholic antecedents, in which the ascending line does not terminate in a plateau, but forms an acute angle with the descending line, which is itself less rigid than in the case of B——.

O

also subarachnoid effusions of blood of the size of a two-franc piece. These slight hæmorrhages appear at the convex surface of the left hemisphere, and in two places on the right hemisphere.

The pia mater is adherent to the cortical layer thoughout nearly the whole of its extent, except in some parts of the occipital lobes and the cerebellum. The adhesions are intimate over the frontal lobes of both sides and the left sphenoidal lobe, where it is impossible to strip off the membrane without taking away with it a thick layer of cortical substance. The basilar artery has a small atheromatous plate, 4 or 5 millimetres ($\frac{1}{5}$ of an inch) wide at its point of bifurcation; some small yellowish plates are also observed on the Sylvian arteries.

The left hemisphere weighs	495 grammes.	7738 grains.
The right hemisphere	570 ,,	8795 ,,
The cerebellum, pons and medulla	155 ,,	2392 ,,
Total	1,220	18,925

Sections made in succession from the periphery to the centre, do not shew any lesions limited to any one point. The floor of the fourth ventricle is strewn with fine granulations, which become nearly confluent towards the point of the calamus scriptorius; the surface of the lateral ventricles is likewise roughened in its anterior half.*

The optic nerves are slightly atrophied, the right smaller than the left, and presenting a greyish gelatinous tint, which is most marked in its under surface, the left also presents a slightly greyish tint throughout its whole extent.

The meninges of the cord are injected, but shew neither false membranes nor hæmorrhages.

Sections of the cord shew no lesion appreciable with the naked eye.

The right lung is the seat of grey hepatization in its lower third; in the middle lobe and at the apex it is greatly engorged : the left lung is engorged at its base; in the larger bronchi of both

* See Magnan and Mierzejewsky, *Des lesions des parois ventriculaires &c.*, (*On the lesions of the walls of the ventricles and the subjacent parts in general paralysis*). [Ependymitis, and diffuse interstitial peri-endymar encephalitis]. *Archives de phys. norm. et path.*, Jan. and March, 1873. See also a paper by Dr. Mierzejewsky on the cerebral lesions in general paralysis, in the same journal for 1875, p. 195.

sides, were found some fragments of chervil and a little curdled milk, (the patient swallowed with much difficulty during the latter days of his life and had often fits of choking after having swallowed mouthfuls of soup, milk or tea).

The pericardium contains about 100 grammes (4 ounces) of turbid serum; the visceral layer is of a milky tint, and presents on the surface of the right ventricle, a whitish patch of the size of a two franc piece. The heart is soft, and loaded with fat on the right side, and is of a brownish yellow tint, which is most marked in the thickest part of the walls of the right ventricle.

The aorta is yellowish on its internal surface and strewn with a great number of atheromatous plates, those most developed not exceeding in size that of a threepenny piece, but for the most part small, prominent, yellowish and resistant; none of them, however, ulcerated. Over the sigmoid valves the fatty degeneration is more marked, and on removing one of the valves, one sees upon the wall of the aorta a yellow line, forming as it were, a hard resistant cord. The valves themselves are pliant and present no signs of fatty infiltration. The liver, yellowish throughout its whole extent, presents in some places portions of a still deeper yellow colour (fatty degeneration). The fibrous capsules of the kidneys are very adherent to the surface, which remains slightly rough after their removal; the whole of the cortical layer is yellowish, and so also are its prolongations between the pyramids.

Portions of the false membranes of the dura mater examined under the microscope shew turgescent vessels, swollen here and there, and presenting at several points small hæmorrhages spreading along the sides of the vessel.

The optic nerves, after remaining for three weeks in a weak solution of chromic acid, were cut into thin sections, some perpendicularly, others in the direction of the axis; in all the preparations, when coloured by carmine, there was seen very marked diffuse interstitial neuritis, principally in the right optic nerve.

The progressive influence of excesses in drinking in the development of general paralysis is here very evident. A first attack of alcoholic delirium is cured in a few weeks, leaving no appreciable traces behind it. After twenty-one months, B—— comes into the asylum a second time in an attack of alco-

holic delirium, after which there appear exalted
ideas of a self-satisfied nature, and the faculties are
evidently enfeebled; this is the commencement of
the period intermediate between alcoholism and
paralysis; the toxic delirium disappears as usual,
but the intelligence remains compromised. Sent
home on trial, B—— is seized at the end of a few
days with an epileptic fit, which is followed by a
delirium of painful character, with hallucinations
and excitement, which necessitates a third confinement. At this time, even during the stage of agitation, intellectual weakness is already noticeable;
the patient moves about in an automatic manner;
he is indifferent, eats gluttonously, picks up things
about him, and deranges his dress. Going out once
more at the request of his wife, he cannot devote
himself to any work, he is excited at times, and has
ambitious ideas, and at the end of eight months he
is brought back, with all the physical and intellectual symptoms of general paralysis.

At the autopsy we have seen the anatomical trans-
lations both of the chronic alcoholic intoxication
and of the general paralysis; the heart, liver, and
kidneys had undergone fatty degeneration, and the
arteries were atheromatous. The heart, moreover,
presented traces of pericarditis, and the fibrous capsule of the kidneys was also thickened and adherent.
As to the nervous centres, we found at the same
time pachymeningitis and diffuse interstitial encephalitis, but the cord was not obviously affected.
These lesions taken together recall in every respect
the anatomical changes obtained experimentally in
dogs by the prolonged action of alcohol; on the
one hand, fatty degeneration and atheroma; and

on the other, chronic inflammation attacking chiefly
the envelopes and interstitial tissue of the organs.

Several other points of interest deserve to arrest our
attention.   The graphic tracing of the pulse showed
us that we had to do with a paralytic alcoholic.
This tracing presents an abrupt ascending line, followed at its summit by a horizontal plateau, after
which the descending line falls again suddenly, all
these being characters which reveal the existence of
atheroma, and a diminution in the elasticity of the
arteries.   It is a tracing analogous to that which the
sphygmograph gives us in chronic alcoholics, and in
cases of senile dementia where the vascular lesions
are of a similar character.   We have thought it right
to place side by side with this for comparison, a tracing of the pulse of a second patient affected with general paralysis, but without any alcoholic antecedents,
in whom, consequently, there, was no atheroma of
the vessels.*

On several different occasions the temperature
shewed an elevation, which each time was related
to apoplectic or epileptiform attacks; in B—— these
symptoms appear to have been of purely cerebral
origin.   The attacks left indeed stupidity and a
very marked obtusion of intellect behind them,
but the physical symptoms were not aggravated;
generally the patient could stand upright, walk,
and display a certain amount of muscular force, the
dementia alone becoming more and more marked.
These symptoms, however, marking out the brain
as the principal seat of the symptoms, agree with
the fact that the autopsy did not discover any lesion
of the cord.   We note this fact, in passing, because

* See Marey, *Physiologie médicale de la circulation du sang.*, Paris,
1863, p. 410, *et seq.*

in other general paralytics, whether the disease be due to alcohol or no, but in whom the cord is more affected than the brain, we see spinal attacks occur. These attacks sometimes shew themselves by numbness and tingling in the limbs, which are usually followed by transient paralysis, (apoplectiform attacks); sometimes by contractions or cramps, and occasionally by convulsions, (epileptiform attacks), without any perceptible modification of the intellect occurring during the fits. But whether the attacks be spinal or cerebral, one symptom which they have in common is never wanting, this is *elevation of temperature.* We have several times had the opportunity of observing these curious phenomena, which in cases of diffuse interstitial encephalo-myelitis bring into play at one time the cord, at another the brain, according to the predominance of the morbid process in one or other part of the cerebro-spinal axis. We reserve for a special work the study of these symptoms, which so often accelerate the progress of general paralysis.

In the last stages of the disease, long after the cessation of all excesses in drink, we see hallucinations of painful character occurring in B——, which in their features recall alcoholic delirium; these were, moreover, associated with ambitious ideas; but we only learn from this that the toxic delirium, even after its disappearance, does not entirely lose its control, and that it may shew itself afresh, at times without appreciable cause, at others, under influences of the most varied character.

ARTICLE V.

Hæmi-anæsthetic form of chronic alcoholism; cases of chronic alcoholism with hemiplegia, and hemi-anæsthesia of the general sensibility and the special senses.

The most frequent mode of termination of chronic

alcoholism, is, as we have already stated, dementia, which is often associated with partial paralysis; this condition answers to fatty degeneration accompanied by atrophy of the cortical layer, together with more or less numerous foci of hæmorrhage or softening, which are generally scattered in the form of lacunar spaces in the optico-striate centres, and in the white matter or at the periphery of the encephalon: all of these being lesions analogous to those of senile dementia, and having their starting point in the changes in the vascular system, that is to say, the atheromatous patches and sometimes miliary aneurisms of the arteries.*

These anatomical lesions, varying in degree and distributed in an irregular manner throughout the vessels of the encephalon, may lead to the most diverse localized affections, some of them of sudden occurrence and rapid course, others, on the contrary, taking place in a slow and progressive manner, or by successive steps; hence arise the numerous symptomatic manifestations, which are variously combined, and often confused with each other both as regards the intellectual phenomena and those of a physical nature, whether motor or sensory. But notwithstanding the complexity of the symptoms, we can nearly always ascribe them to a definite cause by means of a clinical analysis.

Amongst the disturbances of motion and sensation, there appear some symptoms of a special nature, which on more than one account are worthy of our

---

* Bouchereau et Magnan, *Observations d'alcoolisme chronique; hæmorrhagies cérébrales; anévrysmes miliaires du cerveau et de la rétine.* (Compt. rend. et Mém. de la Soc. de Biologie, 1869, p. 36). See also Charcot and Bouchard, *Nouvelles recherches sur la pathogénie de l'hémorrhagie cérébrale, (Arch. de phys. norm. et path.;* 1868, Jan. and Feb., p. 110.)

attention; these are the paralyses of one side of the body accompanied by diminution or total loss of the general sensibility and of the special senses.

After a sudden attack of apoplexy with loss of consciousness, or else, perhaps, by slow degrees, with no other symptoms than headache, numbness, and tingling on one side of the body, the patients observe some muscular weakness; they drag the leg, the hand lets things fall, and sometimes even some thickness of speech, or a deviation of the lines of the face may occur. If we have recourse to the dynamometer we often find marked weakness of the hand, and moreover, the leg is powerless to sustain the weight of the body. Of 10 cases observed at St. Anne, the paralysis was on the right side six times, and on the left, four. The paralyzed arm generally presents a more marked trembling than that of the other side, but in one case only (Case xxx; a patient shewn at the Société de Biologie, on March 16th, 1872), did we observe rhythmic movements at all analogous to those of paralysis agitans, continuing during rest but stopping when the patient was asleep, a kind of trembling to which M. Charcot thinks that some symptomatic significance may be ascribed.*

The whole of the paralysed side presents an anæsthesia, not only of the skin and mucous membranes, but also of the deeper parts. In those cases in which the loss of sensibility is complete, all the various forms of sensation are abolished, touch, tickling, pricking, cold and warm bodies, constant and induced currents, are not felt in any part of the skin over the paralyzed half of the body; the sensibility begins to be excited at one or two centimetres from the

---

* Charcot. *Leçons sur les maladies de système nerveux.* Hystero-epilepsie, vol. i, p. 279.

middle line, and at this level there exists an intermediate zone, which forms a transition between the
anæsthetic and the healthy side; the inter-commu-
nication and the passage of the terminations of the
nerves beyond the middle line, sufficiently explain
this fact.

In the less advanced cases, or when some improvement occurs, the touch, pain, temperature, and
electric stimuli, excite impressions which are easily
perceived, and it becomes difficult to appreciate the
difference between the two sides by these means : we
may, however, obtain some knowledge of the slighter
degrees of anæsthesia by the comparative examination of the sensibility, after Weber's method.
The distance between the two points of the æsthesiometer which is required for the perception of two distinct sensations, will sometimes give a considerable
difference between the two sides of the body. This
separation is, moreover, greater when the points of
the compass are applied in the direction of the axis
of the limb, that is to say, following the direction
of the nerves, than in the case of application across
its axis, when the points are placed on distinct
nerve branches, which answer therefore to two
more widely separated centres of innervation in the
cord. The anæsthesia extends also to the deeper
parts, pressure, in fact, is not felt, and the electro-
muscular sensibility is notably diminished; we may
sometimes pass an induced current through the
muscles of the arm or leg, in such a way as to lead
to strong contractions, almost to tetanize the limb,
without producing pain, whilst the application of
the same current to the healthy limb becomes at
once intolerable. The same phenomena occur with
the constant current. By closing the current on

applying the electrodes, or even whilst they remain
in situ, by alternating the direction of the current
with the commutator (by M. Trouvé's apparatus),
we get contractions in the muscles of the diseased
side, which are nearly always as energetic as those
of the corresponding muscles of the opposite side,
but the electro-muscular sensibility is either *nil* or is
greatly diminished.   The muscular sense is weakened, and sometimes even entirely abolished; when
his eyes are closed, the patient has no consciousness
of his movements, whether spontaneous or excited.
When asked to bring the anæsthetic hand to any
part of his body on the healthy side, he does not
perceive an obstacle interposed between the hand
and the object to be gained, the hand stays motionless or takes a different direction.   But if one
touches the organ named, oneself, for example the
nose or ear on the sound side, the patient thinks
he has executed the movement himself, and says at
once that he is touching it, although his arm has
stopped on the way.   If the part to be reached is
on the anæsthetic side, the patient does not feel
the contact, whether of his own hand or of a foreign
body, unless by a strong pressure he brings into
action the muscles of the healthy side, which warn
him, by their own sensibility, of the resistance experienced by the insensible side.   In walking, when
the eyes are closed, the patient follows a pretty
straight line, but readily allows himself to be drawn
away into walking in a circle, without noticing it,
when one comes and softly holds the anæsthetic
side.   And, lastly, things drop out of their hands
as soon as they cease to watch them, and in various
sorts of work, the patients prick or strike themselves on their fingers without feeling it; thus a

woman who was engaged in sewing, was warned of
her wounds by the drops of blood which spotted
her work. (Case xxxi). The insensible limbs
are habitually colder than their fellows, and in
some cases, the patients themselves have a feeling
of this coldness. Sometimes the difference between
the two hands, when the thermometer is left for a
quarter of an hour *in situ*, may reach two or even
three degrees of the centigrade scale.

The anæsthesia is not confined to the skin, all the
mucous surfaces of the same side are deprived of
sensibility. A quill feather may be drawn over the
palpebral or sclerotic conjunctiva, and sometimes
even over the cornea, without exciting either pain
or movement. In some cases, however, the sensibility is lost over the palpebral and sclerotic conjunctiva, but persists over the cornea, a fact which
is explained by Claude Bernard's researches. M.
Bernard has found that when the ophthalmic ganglion
had been completely removed in dogs, this removal
was always followed by loss of sensibility of the
cornea, with preservation of that of the conjunctiva;
the physiological result thus agreeing with the anatomical arrangement, since the conjunctiva receives
its ciliary nerves direct from the fifth pair, whilst
the cornea is innervated by ciliary branches arising
from the ophthalmic ganglion. On the other hand,
poisonous substances exercise a different action on
these two parts; in death from strychnia, the cornea
is the first to become insensible, whilst in anæsthesia
from ether and poisoning by curare, it is the conjunctiva which first loses its sensibility.* A phenomenon
which we have observed in two of our patients, and

* Cl. Bernard. *Effets des substances toxiques et médicamenteuses.*
Paris, 1857, p. 235.

which Demeaux* also found in a woman affected
by hemiplegia with anæsthesia, is the appearance of
tears in both eyes on the contact of a foreign body
with the cornea of the insensible side, without either
the patient's having felt it, or the globe of the eye
making the slightest movement; the insensibility of
the cornea is complete, and yet the reflex action is
shewn by the immediate secretion of tears. In our
two cases the palpebral and sclerotic conjunctiva
were equally insensible; but irritation by touching
them was not followed by a flow of tears. We may
in some degree liken this fact to what occurs in
hysterical hemi-anæsthesia; the erectile tissues, as
Briquet says, even although they do not transmit
the impression to the brain, retain their power of
erection; thus the nipples of the breast and the
clitoris, although insensible to the touch, retain
their faculty of erection on the slightest contact.†

The exploration of the nasal fossæ, and of the
external auditory meatus, shew in like manner a
diminution or complete abolition of the sensibility
of their mucous membrane. The buccal cavity
and the tongue are also the seat of hemi-anæsthesia, whether the sensations successively called
into play be those of temperature, pain, or simple
contact. Tickling the corresponding half of the
soft palate or of the uvula does not excite any reflex
nausea; striking the teeth is not felt, although
sometimes the patient hears it with the healthy
ear, or perceives it in the non-anæsthetic side at

* Demeaux. *Des hernies crurales*, (Thèse de Paris, 1843. The
case of hemiplegia with hemi-anæsthesia is at the end of the
essay).

† Briquet. *Traité Clinique et therapeutique de l'hysterie.* Paris,
1859, p. 283.

the tempero-maxillary articulation, by the vibrations or the shaking produced in the lower jaw. The anæsthesia also exists, on the affected side, on the mucous membrane of the glans, the meatus urinarius, and the margin of the anus.

The special senses present not less remarkable modifications, which vary in degree from simple weakening to a complete abolition of function.

It is interesting to follow these modifications of function in each of the senses, whether the affection increases in a progressive manner, or, on the contrary, after a sudden loss of the special sense, the normal functional condition is regained.

The sight may be the seat not only of weakening, but in some cases of a peculiar perversion which is also found in other cases of simple alcoholism. When the acuteness of vision is but slightly impaired, a comparison between the two eyes by the typographic scale indicates slight differences. Characters of 2 millimetres in size are easily read by the healthy eye, whilst the affected eye can only distinguish letters of 1 or 2 millimetres, and the difference is usually even greater than this; often, too, the characters speedily become surrounded by a mist, and reading cannot be continued for more than a few seconds. In a higher degree of amblyopia, both persons and objects cease to be recognized, and finally a degree is reached in which light is not distinguished from darkness, and the sight on that side is completely abolished. In the patient in case xxix, exposure to the rays of the sun left the eye open and motionless, although the pupil became slightly contracted; mobility, however, was retained, notwithstanding the amaurosis; and moreover it was only necessary slightly to lower or raise the eyelid to obtain the usual movements of contraction and dilatation of the pupil; slight dilatation occurring when the lids were closed, and contraction immediately on opening them. But such is not the case in all hemi-anæsthetics, and in his case, Demeaux notes the occurrence of immobility of the pupil. In hysterical hemi-anæsthetics affected with amblyopia, Briquet also mentions immobility of the pupil as in ordinary cases of amaurosis.

The passage of a continued current through the head or its neighbourhood does not produce *phosphènes*, (flashes of light), in the anæsthetic eye, whilst the healthy eye perceives them on opening or closing the current.

In addition to amblyopia, there is in some of these patients a unilateral dyschromatopsy; the healthy eye distinguishes all colours rapidly and completely, whilst the anæsthetic eye confuses not only the composite colours, but pure and deep colours too.

In one of our cases of hemi-anæsthesia, the dyschromatopsy was bilateral, just as it is in some other alcoholics, although much more rarely than M. Galezowski states.

The ophthalmoscopic examination does not reveal any special lesion of the fundus oculi, even in patients whose sight is entirely abolished; we may notice venous stasis, and sometimes perivascular infiltration around the disc, and even pulsation in the veins may appear; but no important lesion of the papilla or retina is found to explain the unilateral loss of sight.

The hearing is weakened, the ticking of a watch is heard at a distance of 5 to 10 centimetres, (2 to 4 inches), on the diseased side, whilst the other ear perceives them at 50 to 60 centimetres, (20 to 24 inches), as in the normal condition. In some cases the deafness is complete, neither the voice, nor any musical note, nor any noise being heard on the diseased side. On placing the button of the positive pole over the mastoid region of the healthy side, and applying the end of the other rheophore, reduced to a simple wire, to a plug of cotton wool introduced into the external auditory meatus of the anæsthetic ear, closure of the continuous current does not produce any sensation of sound; whilst the application of the same excitant in the contrary direction produces a distinct perception of sound in the healthy ear. There is sometimes a metallic taste at the same moment on the corresponding half of the tongue.*

The sense of smell is also weakened, and may in some cases be completely abolished. In order to appreciate the degree of anosmia, it is necessary to employ organoleptic excitants of varying strength. Thus we may graduate the stimulant action by employing in succession, orange-flower water, camphor, essence of peppermint, tincture of musk, vinegar, and essence of mustard. These two latter agents address themselves at once to the common and to the special sensibility, and in some cases only a simple irritation is produced, and the patients' tears flow : or they feel pricking sensations, but do not perceive any odour.

Taste is diminished or lost on the anæsthetic side; we may discover this by the aid of substances of progressively more pronounced taste, such as sugar,

---

* For an explanation of a parallel fact, see Wilde's *Aural Surgery*, p. 305. (*Trans. note*).

salt, sulphate of magnesia, aloes, colocynth, and
pimento. Or we may employ solutions of varying
degrees of concentration, or else simply hold the
sapid body on the tongue, beginning with the anæsthetic side, and if, at the end of one or two minutes,
the patient indicates by a movement of the head
that he does not taste it, the substance may be carried to the healthy side of the organ, and its taste
will be immediately recognized.

The application of the continuous current to the
anæsthetic side does not excite any gustatory sensation; whilst on the other side, it gives rise to a
somewhat marked styptic metallic taste. At the
moment of closing the current, the muscles contract
on the side on which the poles are placed, but the
patient does not feel the movements except on the
healthy side of the organ, just as is the case in
the other muscles of the corresponding half of the
body.

In hysterical persons, the hemi-anæsthesia is by far
most frequently on the left side; whilst in chronic
alcoholism both sides, as we have seen, are indiffer-
ently affected, and in our ten cases, eight of whom
were men and two women, the hemi-anæsthesia
with paralysis was on the right side six times, and
on the left four. Of the two women, one was
hemi-anæsthetic on the right side, and the other
on the left.

In this hemiplegia of sensation and movement,
the disturbances of the motor functions sometimes
improve rapidly, whilst the anæsthesia persists; at
other times both kinds of symptoms disappear together; and in yet other cases, as we see in two of
the patients, motion and sensation tend to become
progressively enfeebled. In one of these, however,

there was already observed a complete abolition of the general and special sensibility of the whole right side, whilst on the other hand, although the power of movement was weakened, the paralysis was far from having reached so advanced a degree as the anæsthesia.

These patients, like other chronic alcoholics, suffer from headache, attacks of giddiness, humming in the ears, numbness and cramps in the limbs, sometimes more marked on the paralyzed side, but rarely limited to it; but in them attacks of vertigo, apoplectiform and epileptiform seizures shew themselves with greater frequency. This is owing to the fact that in addition to the diffuse or disseminated lesions proper to chronic alcoholics, these hemiplegic subjects are also under the influence of a localized lesion, a sort of thorn or irritating agent, whose action is superadded to the disturbances produced by the atheromatous degeneration of the vessels.

The intellectual faculties are also enfeebled, the memory is weakened, the judgment less clear, and imagination extinguished; the ideas, confused and ill-coordinated, are incoherent in their expression, the moral sense is greatly weakened, and a maudlin sentimental condition becomes added to the indifference and apathy. The delirium, although losing its activity, retains nevertheless its general fundamental characters, and is of the disagreeable and painful character which it presents in acute alcoholism. Illusions and hallucinations, if they still exist, have no longer, with some exceptions, that vividness and remarkable multiplicity which they possess in the earlier periods of alcoholism, but the sleep is still disturbed by dreams and nightmares, hypochondriacal notions often supervene, with

melancholic ideas and sometimes even a suicidal tendency; but the delirious outbreaks are
less acute, the reaction is less sudden, less noisy,
and there is a sort of uncertainty and indecision
reigning over all these illusory perceptions of the
mind.   As we see, but for the hemi-anæsthesia
there is nothing in the symptoms peculiar to this
form of chronic alcoholism.   Moreover, these singular disturbances of sensation do not appear solely in
alcoholism, and have not an exclusive relation either
with a hæmorrhage, a patch of softening or of sclerosis; either of these lesions may give rise to them, but
only on condition that the structural change affects
a determinate point of the encephalon; it is, then,
rather a question of the seat or topography of the
lesion, than of its particular character.   On the other
hand we must note that in the female in hysteria,
this group of symptoms may be only a more or less
transient simple functional disturbance, without any
appreciable material lesion.   In order to assign
to the different symptoms their true value, it is
necessary to determine what are the regions of the
nervous centres which preside over the free exercise
of general and special sensation.   According to a
theory admitted by several physiologists in France,
it is believed that the centre for the perception of sensory impressions is situated in the pons Varolii; this
was the opinion of Longet, and is now the opinion of
M. Vulpian.*   The experiments are well known;

* Longet, *Anatomie et physiologie du système nerveux.* Paris,
1842, vol. 1, p. 431.—Vulpian, *Physiologie du système nerveux.*
Paris, 1866, p. 543.—According to M. Béclard, the pons is a conductor of sensory and motor impressions in the same way as
the cord and medulla oblongata.   *Traité de Physiologie.*   Paris,
1862, p. 997.

the cerebral hemispheres and the cerebellum are re-moved in a rabbit, leaving the pons entire; peripheral irritation, such as pinching the nose or ear, excites plaintive cries, cries of acute pain and not merely of reflex origin. On the other hand the animal scratches its nose several times with the fore-paws after smelling ammonia. In the grey field mouse similarly deprived of the cerebral hemispheres, corpora striata and thalami optici, pinching the ear excites a general movement of the limbs with extension of the head. After blowing on the ear suddenly the animal shakes its head and ears, and winks the eyes, thus adapting the reaction to the nature of the irritant. The centre for the percep-tion of sensory impressions is therefore situated in the pons. M. Vulpian adds with regard to the pons, that it seems to preside not only over the general sensibility but also over certain special senses. " Thus it seems to me certain," says this author, " that the auditory and gustatory sensations take place in this part of the nervous centres. In the rat deprived of the cerebral hemispheres, the corpora striata and thalami optici, the slightest noise makes it tremble, and the sudden puff imitating that which cats make when angry causes it to make a sudden leap." Whilst admitting with M. Vulpian the influence of the pons on hearing and taste, a lesion of this centre alone would not explain the loss of smell and sight which exists in most of the patients in question.

According to another theory, that of Todd and Carpenter, the centre of perception of tactile im-pressions would be situated higher up, and reside in the thalamus opticus, at which point the sensory impressions would be transformed into sensations;

on the other hand, the corpus striatum terminating the motor tract, would stand in a similar relation to the execution of voluntary movement.

It is evident that it is only by the careful investigation of such facts as those which we are now considering, that we can hope to elucidate these difficult questions of pathological physiology. But the clinical and pathological results already obtained, may serve as our guide, and direct our attention to the regions which appear to be those which must be the seat of these changes. To understand most of these observations, especially those of L. Turck, and to follow advantageously the distribution of the lesions which he exhibits in the plates which follow his memoir, it is necessary carefully to examine and to understand the reciprocal relations of the principal groups of cells or ganglionic-centres seen on a transverse section of the brain, made immediately behind the corpora albicantia. In the median line, going from above downwards, we see in this section, the corpus callosum, the velum interpositum, and the third ventricle with the grey commissure, (commissura mollis). On either side and close to the middle line is the thalamus opticus, whilst further out and above is the nucleus caudatus or intra-ventricular nucleus of the corpus striatum; along the optic thalamus runs the internal capsule, formed by the radiating portion of the base of the cerebral peduncle, and by fibres which pass from the white centre of the hemisphere to the optic thalamus; still further appears the extra-ventricular nucleus of the corpus striatum or nucleus lenticularis, which is subdivided into three concentric segments. Between the intra- and extra-ventricular nuclei of the corpus striatum,

is seen the apparent continuation of the internal
capsule, or origin of the great corona radiata of Reil,
which, according to Wagner and Kölliker, Luys
and other anatomists, is not the prolongation of the
cerebral peduncles, but rather the re-union of fibres,
which, arising from the white centre of the hemisphere, or rather from its cortical layer, run towards
the optic thalamus, where, for the most part, they terminate. Outside the lenticular nucleus appear, 1, a
narrow white band or external capsule, 2, a grey band,
the nucleus tæniæformis or *claustrum*, and lastly, 3,
the island of Reil. To resume, from within outwards,
we see; 1, the thalamus opticus, and above it the intra-ventricular nucleus of the corpus striatum; 2, the
internal capsule with the corona radiata of Reil; 3,
the intra-ventricular nucleus, and further out the external capsule and claustrum (*avantmur*); and 4, the
insula.

Of the whole of this wide region, one part seems
to be the special seat of the lesions which are
accompanied with hemi-anæsthesia of general sensation, and of the special senses ; this is a sort of cross-
way which is in relation with the different centres,
namely, the space included by the superior extremity of the internal capsule, the origin of the corona
radiata, and the upper portion of the lenticular nucleus. In the cases recorded by M. Charcot, of paralysis with hemi-anæsthesia and trembling of the
same side of the body, the lesion occupied the posterior region of the thalamus opticus and the ad-
jacent parts of the cerebral hemisphere situated
external to it.

The thalami optici and the parts bordering on their
outer side have been involved in some other cases
of hemi-anæsthesia of the general sensibility cited

by various authors; but the only cases followed by an autopsy in which, in addition to anæsthesia of common sensation, there was any note of loss of special sensibility, are those of L. Turck.* In the four cases of Turck, the hemiplegia was in two cases accompanied by hemi-anæsthesia of general sensation, with loss of the senses of taste, smell and sight; in one of these cases, the lesion consisted in a very extensive softening of the white centre of the left hemisphere, and included the outer two thirds of the corona radiata of Reil. In the other case, it was a tract of hæmorrhage, situated in the outer part of the left optic thalamus, involving also a part of the internal capsule.

In the other two cases, where the general sensation alone was compromised, the lesions consisted, in the first case, of three patches of softening in the right hemisphere, occupying, but with intervals of healthy tissue, the space comprised between the nucleus caudatus and the lenticular nucleus; one of these patches was situated immediately outside the tail of the nucleus caudatus; another was found external and posterior to the first, on the course of the corona radiata; the third occupied a part of the external segment of the lenticular nucleus, extending a little beyond it at the upper part, so that the corona radiata was separated in places from the internal capsule and the optic thalamus.

In the second case there were two old hæmorrhagic foci in the right hemisphere, one situated on the upper and outer margin of the optic thalamus,

* Ludwig Türck, *Ueber die Beziehung gewisser Krankheitsherde des grossen gehirnes zur anästhesie. (Sitzungsberichte der mathem. natur. Classe der kais. Akad. der Wissenschaften*, 1859, Band xxxvi., § 191).

on which it slightly encroached, the other in the third segment or external portion of the lenticular nucleus of the corpus striatum.

To sum up then, the upper and outer part of the optic thalamus, the lenticular nucleus, the internal capsule and corona radiata, these are the parts affected; but it must be observed that in all the cases, even when the lesion was situated in the lenticular nucleus of the corpus striatum, it exceeded the limits of this, and involved in some measure the corona radiata, which in fact is the region most frequently affected. This region was also affected in a recent case, which M. Charcot has been kind enough to communicate to me; it was a case of hemi-anæsthesia with choreiform movements of the right side, and at the autopsy there was found a lacunar patch of softening on the left side at the foot of the corona radiata in the convolution called the *operculum*.

The following cases added to the two already related (Cases xix and xxiii) of paralysis with hemi-anæsthesia of the general sensibility and of the senses, will give a pretty clear idea of the characters and course of these symptoms.

CASE XXIX.—*Excesses in drinking; absinthe.—Headache; vertigo; epileptic attacks.—Trembling; hallucinations.—Apoplectic attack; incomplete right hemiplegia; right hemi-anæsthesia with diminution, and subsequently loss, of taste and smell; impairment of hearing followed by complete deafness; amblyopia and dyschromatopsy terminating in total loss of sight of the right eye.*

Pierre L—— aged 59, a day labourer, formerly a soldier, who for some years had been addicted to alcoholic drinks, contracted whilst in Africa the habit of drinking absinthe, of which he took as much as six or seven glasses a day. Shortly after these excesses, in addition to suffering from habitual sleeplessness with dreams and nightmare, loss of appetite and bringing up phlegm in the morning, there were superadded violent attacks of headache, dizziness, giddiness, and convulsive seizures, accompanied by loss of consciousness, biting of the tongue, and involuntary evacuations. Still later, he complained of numbness, tingling and cramps in the limbs. On repeated occasions there was observed trembling of the hands, he was tormented by frightful visions, and he heard insults and threats against his life. This condition continued for several years, becoming better or worse according as L—— had greater or less opportunities for indulging in his drunken habits. During a period of four years, from 1865 to 1869, he took long sea voyages, he became more sober, and his health improved; then he landed at Martinique and stayed at St. Pierre for some time, where he began to drink again. One day, whilst at work, he suddenly lost consciousness for some time, how long he cannot exactly tell, but he remembers that when he came to his senses he found himself in a hospital, his speech was much embarrassed, and he was paralysed on the right side.

Six weeks after this occurrence, the muscular weakness diminished, and he was able to get up and walk, dragging his right leg; at the end of four months, they were able to employ him as a hospital attendant, giving him, however, the lightest work. In 1871, he came back to France, and on his return to Paris, he lived wretchedly, with difficulty earning enough to provide for the absolute necessaries of life, but, nevertheless, not giving up his drinking habits.

On his admission to the asylum on June 12th, 1872, the following notes were taken. He appears stupified, hardly understands questions addressed to him, turns round and looks about him in a frightened way; he hears himself called names; he sees cats, ghosts, black men. He complains of headache, dizziness and vertigo; he has trembling of the hands which is most marked in the right, and there is weakness of the whole right side of the body. During the night he is disturbed by dreams, nightmares and hallucinations; in the day time he becomes calmer, but still appears restless and preoccupied.

June 14th to 30th. He is gaining confidence by degrees; no longer suffers from hallucinations during the day, but he seems apathetic and indifferent, and when not spoken to remains seated with his arms crossed; he works and plays when he is pressed to do so, but he does nothing spontaneously and does not complain at all of his confinement; his faculties are moreover enfeebled, and bis memory treacherous. From time to time during the night, he sees animals which terrify him, apes, horses and hyænas.

He complains constantly of heaviness of the head, humming in the ears, dazzling sensations, flashes of light and flames before his eyes, and cramps in the limbs, chiefly in the right side. The trembling is more marked in the right hand, especially when the upper limbs are held stretched out for a minute or two : fatigue soon occurring, the right arm tends to fall, and to the increased trembling, there are added little irregular shakings in the fingers. In the first exploration with the dynamometer, the right hand presses the index to the 16th division, and the left to the 27th. This examination repeated from day to day always indicates a greater weakness of the right hand, which reaches to 22°-20°-and 17,° whilst the left hand carries the index to 31°-37°-and 31.° The right leg is also the weaker, the patient can jump with the left foot, but the right cannot be raised from the ground. The paralysed side is the seat of notable anæsthesia ; touch, tickling, pricking, and cold and warm bodies are hardly felt, and this is the case, not only in the arm and leg, but also on the right side of the head and trunk.

The sense of smell is very much diminished in the right nostril ; orange flower water, camphor and essence of peppermint arouse no sensation, vinegar slightly affects the mucous membrane, but the patient does not recognise it. The nature of these substances is well appreciated by the left nostril.

The right half of the tongue does not perceive either sugar, salt or colocynth, the left half does distinguish these substances. The buccal mucous membrane is anæsthetic on the right side and feels but very imperfectly either touch, pricking, or the action of heat or cold.

The vision is less acute on the right side ; the right eye can only read letters of 1 centimetre in diameter, the left reads characters of 3 millimetres.

The right eye is affected with dyschromatopsy, it distinguishes the colours red and blue, (No. 10 of Dr. Galezowski's chromatic scale), but violet appears white, and yellow whitish. The left eye recognizes all the colours at the first glance.

July. The intellectual condition remains stationary, the patient appears completely indifferent ; he repeats the same things, sometimes contradicts himself, gives wrong dates ; at times he is in a maudlin condition and cries without reason. The trembling has much diminished, it is only necessary, however, to make him stretch his arms out to cause it to reappear in the right hand. The paralysis of movement and of the general sensibility and the senses has not increased. The right leg quickly becomes numb when the patient remains seated, it gives way as soon as he gets up, and he is obliged to raise it with his hands, shake it and stretch it on the ground to make walking possible.

He frequently suffers from headache, a sensation of beating in the frontal region, and attacks of giddiness; cramps and tingling, especially on the right side.

August. Intellectual condition the same, slowness of ideas, still has nightmares and hallucinations; he sees phantoms, and people talking, and hears people fidgetting about his bed.

Frequent attacks of dizziness and vertigo, mists before the eyes, shooting sensations and cramps in the right arm and leg. The cutaneous anæsthesia has not made any progress, rather strong pricking and painful sensations are still perceived. On application of the two points of the æsthesiometer in the direction of the axis of the limb, on the outer surface of the right thigh, there is a sensation as of one prick only when the points are 13 centimetres (5 inches) apart; the sensation of two pricks being obtained with a separation of 15 cent. (6 inches). On the left side, in the corresponding situation, the patient feels two pricks at the distance of 9 centimetres.

On the inner aspect of the right thigh, the sensation of one prick is obtained by a distance of 10 cent. between the two points; the sensation of two pricks is produced by increasing the separation 1 or 2 cent.; on the left side and in the same region the points of the æsthesiometer when 6 cent. apart, give the sensation of two punctures. On the outer aspect of the right arm the application of two points at 10 cent. apart gives the idea of a single prick, on the left arm a separation of 6 cent. allows the double sensation to be felt. The right side of the forehead only feels one prick, the two points being six cent. apart, whilst on the left, the sensation of two points is obtained at a distance of only 3 cent.

September and October. The patient continues to experience attacks of dizziness and vertigo, accompanied by alternate pallor and flushing of the face, numbness, tingling and cramps, sometimes also contraction of the arm and leg of the right side with stiffness and flexion of the fingers, which contract again after straightening. At times, too, there comes on a pain in the right cheek, with slight tremors, or rather an involuntary snapping of the jaws. Orange flower water, camphor, essence of mint and tincture of musk provoke no sensation in the right nostril, but are recognized with the left.

Sugar, salt, and colocynth produce no recognition of taste on the right side of the tongue ; on the left, on the contrary, their characters are perfectly appreciated.

The right ear, which is the seat of incessant humming noises, hears the ticking of a watch only at the distance of about an inch from the external ear; on the left side, on the other hand, they are heard at 12 inches.

November and December. Frequent vertigo, dizziness, loss of appetite, pallor, altered countenance, and expression of suffering. Sometimes in the evening, a slight febrile attack and increased headache. Examination of the chest shews nothing noteworthy. Urine clear, amber coloured, and, containing neither sugar nor albumen. From time to time, hallucinations during the night; L—— gets up, looks under the bed and in the corners of the room, and says he thinks some one has come into the room and is calling him, and has been dragging him out of bed by the feet.

On the right side the hemiplegia is increasing, and there are painful cramps in the arm and leg; the cutaneous anæsthesia is progressing, and pins stuck deeply into the skin are hardly felt, and the patient localizes the pain in a situation nearer the trunk; thus he refers to the leg a touch on the foot, to the knee one on the calf, and to the thigh that on the knee.

The sense of smell is almost *nil* on the right side; acetic acid held under the nostril, although it excites a flow of tears, is hardly felt; on the left side all odours continue to be precisely estimated.

No taste is recognized on the right side. The hearing grows progressively weaker, and the watch must be placed upon the ear itself for its ticking to be heard ; the humming sounds are frequent; at times there are tinkling sounds and noises as of a bell ringing, and the patient says that it seems as if the air constantly got into his ears.

The sight is becoming dim on the right side, all objects appear as if enveloped in a cloud, and even the largest letters cannot be distinguished. Ophthalmoscopic examination, repeated several times, discovers nothing wrong either in the media or the membranes of the eye; the papilla is physiological, of rosy tint, the vessels normal, all the parts of the fundus appear normal, and there is nothing particular to be seen in the region of the macula.

Jan. and Feb., 1873. The faculties continue to become weaker, the giddiness and vertigo are sometimes associated with a stupidity which lasts the whole day. Sleep is disturbed by nightmares and painful hallucinations. L —— sees fire, and incendiaries, hears uproars, drumming and the firing of volleys.

The muscular power is diminished on the right side ; the anæsthesia is very marked, and pinching or pricking with a pin is no longer felt; when his eyes are shut, the patient has no consciousness of passive movements of the right arm and leg ; he no longer feels either cold or warm bodies, nor heavy or light weights laid upon the skin.

Hummings and ringing noises in the right ear; the voice, even when very loud, is hardly heard. On the same side vision continues to grow weaker, and taste is lost; the right nostril cannot distinguish any smell. On the left side, general sensibility is not modified, and the organs of sense are intact.

March, April, and May. The faculties are weakened, but the dementia appears more profound than it really is, because of the stupor accompanying the attacks of dizziness and vertigo. Dreams and nightmares continue to disturb the sleep. The arm and leg of the right side are subject to numbness, tremor and cramps, with stiffness of the fingers, the paralysis is in a stationary condition, hemianæsthesia is complete ; the muscular sense is completely abolished. If, when the eyes are shut, one checks the right arm whilst performing a movement, L —— does not know whether the movement is completed ; and when the object to be reached is a part of the healthy side of the body, such as the cheek or ear, as soon as these are touched by some one else's hand, L —— thinks he effected the movement and says that he has touched the part indicated. The anæsthesia of the special senses on the right side persists, on the other side their sensations are normal.

June, July. Attacks of vertigo, heaviness of the head, throbbing in the forehead, humming noises in the ear, most marked on the right side, and hallucinations, especially at night.

The right hemiplegia increases slightly, the patient cannot sustain himself on the right foot even when he supports himself by leaning against the furniture ; he suffers from tremors, numbness, and cramps in the arm and leg.

Anæsthesia is complete throughout the whole right side to within two centimetres of the median line, where sensation begins to reappear, becoming normal on the left side at one or two centimetres beyond the median plane. On the trunk, the arm, leg and face, a pin can be passed through the skin without the patient feeling the slightest sensation. The same is the case with buccal, nasal and palpebral, and other mucous surfaces; a spoon slowly drawn over the right side of the velum palati does not give rise to retching; whether heated or cooled, its contact is not felt over any part of the right half of the buccal cavity. Foreign bodies introduced into the right nostril or the external auditory meatus do not excite any sensation.

The patient has no knowledge of the movements of the right hand when the eye does not direct them.

The arm and leg of the right side are cooler to the touch than those of the left side, and the thermometer kept for a quarter of an hour in each hand perfectly closed rises to 34·8° C. (94·6° F.) in the left, and 34·2° C. (94·6° F.) in the right. The anæsthesia of the organs of sense has not diminished.

The exploration of the sensibility of the superficial and deep parts by the aid of the continuous current with 50 elements (of Trouvé's apparatus) gave the following results:—The current is not felt in any part of the right side. On moistening the conductors, and using the commutator fitted to the apparatus, we get, on closing or opening the current, sudden contractions of each group of muscles, equally energetic on the right as on the left side, but on the right side the contraction is not attended with any sensation, and the patient is not aware of the actions which take place except only by seeing the contraction of the muscles, whilst on the left side this is accompanied with severe pain.

The application of the two poles on each side of the nape of the neck or to the two temples causes *phosphènes.* With 7 elements, the patient perceives, he says, gleams of light, with 9 to 15 elements, these are like flashes of lightning, but this perception of light only occurs in the left eye, the right remaining entirely insensible.

The two poles applied on the right side of the tongue with 20, 30, and 50 elements provoke, with the aid of the commutator, abrupt contractions in the muscles of the right side, but do not give rise to any sensation; on the left side on the contrary, the application of the conductors to the tongue with 20 elements excite painful contractions, with a very marked metallic taste, and a sensation of pricking and burning.

When the negative pole is applied over the right mastoid process, and the plain wire of the positive pole is placed upon a plug of moistened cotton introduced into the left external auditory meatus, it produces with 15 elements the sensation of a sound, which according to the patient, is like the tinkling of crystal. On applying the negative conductor over the left mastoid process, and the other on the wetted cotton in the right ear, no sound is perceived, whatever the intensity of the current, with 20, 30, or 50 cells.

The application of the induced current on the right side of the body does not give rise to any sensation, whether the skin is electrized by means of the electric brush, or whether the current penetrates deeply by the aid of a pad; if they are wetted the muscular contractions under the influence of slow or rapid interruptions, and whatever the intensity of the current, are not felt, although the contractions may be as energetic as in the normal condition. On the left side the electro-cutaneous sensibility is intact, as is also the electro-muscular sensibility.

August—September. The intellectual enfeeblement does not shew any noteable change. Anæsthesia is complete on the right side; one day the patient burnt the back of his hand against the plate of a furnace without feeling it. Another time he placed his right leg in a foot bath which was too hot, and left it there until on putting the left foot in its turn, into the bath, he was warned by it of the high temperature of the liquid. The deeper sensibility is similarly abolished. On the other hand, the right hand cannot effectually perform any movement, unless it is directed by observation. In walking with the eyes shut, it is only necessary to press slighly on the anæsthetic side, to make L——, unknown to himself, walk round in a circle. The paralysis continues to undergo an increase, the right foot drags slightly on the ground, the right hand presses the index of the dynamometer to the sixth division, the left reaching 25°. The whole right side is cooler than the left, and by means of a comparative examination repeated several times, a notable difference is discovered, always to the prejudice of the anæsthetic hand. On August 11th, the right hand gives 33·8° C (93° F.) the left 36° (96·8° F.); on the 12th, the right hand 27·8° (82° F.), the left 30·8° (87·4° F.); on the 13th, the right hand 32·6° (90·5° F.), the left 34·1° (93·4° F.); on the 15th, the right hand 29·5° (85° F.), the left, 30·6° (87° F.) Special sensation is abolished in the right side; the right nostril not perceiving any odour, and a bottle of acetic acid placed under it is not felt, notwithstanding its combined odorous and exciting properties; the latter, however, exerts some action on the eyes, which become filled with tears. Colocynth and pimento are without action on the right side of the tongue. The right ear hears nothing now; blindness is complete on the right side ; the eye remains open to the sun without perceiving any sensation, notwithstanding that the mobility of the pupil persists, it dilates in the shade, and contracts to light.

October. For some days L—— has been suffering from headaches, more frequent attacks of giddiness, gastric disturbance and fever in the evening. The delirium and the painful hallucinations have shown themselves afresh. L—— saw flames, fires, and imagined that he had thrown children out of the window. At the end of six days these symptoms ceased, sleep and appetite improved, and the patient fell again into his usual apathy. Exploration of the general sensibility and the senses, both by organoleptic agents and continuous and induced currents give results similar to those of the last month ; hemi-anæsthesia is complete.

The right eye, which is blind, when examined again with the ophthalmoscope, shews nothing abnormal ; the papilla preserves its transparence and physiological rosy tint, the vessels, both arteries and veins, pursue their regular course across without any particular alteration either at or beyond the papilla.

The pulse, of which we give a tracing taken from the left radial, reveals the profound alteration which the vascular system has already undergone ; the ascending line is followed, in fact, by a very wide plateau, after which the descending line falls abruptly. These as we know are all characters which are assigned to atheroma of vessels.

Trace III.—Pulse of L—— Left radial; Nov. 15th. 1873. Chronic Alcoholism.

The treatment, apart from some saline and alcoholic purgatives, consisted in the employment of iodide of potassium and tonics, wine and extract of quinine, bitter infusion and sulphur baths.

This case possesses the advantage of showing the gradual progress of the disturbances of sensation, if not from the very beginning of the symptoms, at least from the period when the general sensibility began to be blunted and the special sensibility to become perverted and weakened. The disturbances of vision terminating in complete blindness, without the slightest' appreciable change in the papilla or retina, demonstrate the exclusively cerebral seat (*i.e.* limited to the hemisphere) of the material morbid change. Lesions of the corpora quadrigemina and those of the base of the brain, or even lesions which though further removed act by compression, are accompanied pretty rapidly by changes in the optic tracts, chiasma, and optic nerves, which soon show themselves by manifest alterations of the fundus oculi. It is probable that the other nerves of special sensation, notwithstanding their loss of function, are not further altered, and the anatomical determination of this fact will not be without importance, from the point of view of the pathological physiology of the brain.

Lastly, there are two special points to which we may recur apart from what has been said in the course of this account, viz., the appearance in this patient, at the very outset, of epileptic attacks following the abuse of absinthe (absinthic epilepsy); whilst at the end of the intoxication we find attacks of giddiness and apoplectiform seizures predominating, symptomatic of material lesions already produced in the brain. In the second place, L—— has on several occasions, without further excesses, but under the influence of illhealth or fever, presented transient attacks of delirium of a painful character entirely resembling alcoholic delirium.

CASE XXX.—*Drinking excesses from the age of* 19. *Abuse of absinthe (epileptic attack).—Defective hearing and paresis on the right side for two years; trembling more marked on the right side.—*

Attack of alcoholic delirium.—Hemianæsthesia and muscular weakness on the right side.—Amblyopia.—Defective hearing, with abolition of taste and smell on the right side.—Marked improvement.

Francois M—— æt. 37 years, a drayman, came into the St. Anne on March 7th, 1873, with the following medical certificate: " Acute alcoholic delirium; incoherent rambling; attacks of terror; suicidal attempts. The patient says that they want to take out his brains."

During his infancy, M—— fractured his right forearm, but his general health has been good; enrolled as a conscript at the age of 19, he speedily contracted the habit of drinking. During the campaigns in the Crimea and in Italy, in the latter of which he was wounded in the thigh and had one toe frost-bitten, he several times suffered from trembling of the hands, a sense of weight in the head, and pains in the stomach accompanied by expectoration of phlegm in the morning, which was as he expressed it *bitter as gall*.

In the last two years passed under the colours, he addicted himself more especially to absinthe, and he appears to have had an attack of alcoholic delirium with convulsive symptoms of which he cannot furnish any precise account; he seems however to have bitten his tongue and passed water in his bed.

Married at the age of 30, he has had four children; one ricketty who died early; a second succumbed to the sequelæ of an attack of broncho-pneumonia; the two who are living enjoy good health.

He declares that he never had syphilis, and in fact, no traces of any such disease are found on his body.

Five years ago, he was caught between the buffers of two railway waggons, and dragged along the railway for five or six yards; the pressure was exerted on the abdomen and the lumbar region. For three months he could not hold himself upright, then he began to walk with the aid of a stick, and at the end of six months he was able to resume his work, but he remained rather weak for some time longer.

About the same time, he suffered from giddiness, everything seemed to turn round him, and he was obliged to stop, and to support himself or sit down lest he should fall. For two years past he noticed that the right ear was hard of hearing; and since last year the arm and leg of the right side have begun to tremble; and with this peculiarity, which was repeated several times, that the trembling was most marked on *Monday*, that is, on the day following that on which his drinking excesses were greatest. He was then drinking beer and also spirits, especially gin.

In the early part of March, after fresh potations, he suddenly quitted his wife and children, and set out from Lille for Paris, with the money which he had collected during the day on his master's account. At Paris, he indulged in an unbridled debauch for several days.

Seized with an attack of alcoholic delirium, he hangs half out of the window of his room, and remains in this attitude crying and groaning, not daring to leap into the street, where he says, an immense crowd want to throw him into the water, nor to return into the room, which is occupied by armed men who want to assassinate him. He is overpowered, and taken before the Commissary of police; whilst there, seizing a knife, he tries to commit suicide, and makes two stabs in the præcordial region.

On his entrance to the Asylum, he is a prey to incessant hallucinations. People threaten him, insult him and want to assassinate him; he sees men armed with sabres prowling about him; he sees policemen.

There is considerable trembling of the arms and legs, most marked on the right side. Tongue moist; no fever; urine contains neither sugar nor albumen. Treatment; two bottles of citric lemonade; a jug of broth, and a bath.

March 11th. The first night was passed without sleep; next day the patient was quiet for some hours in the middle of the day. This morning, he answered questions pretty exactly, and he has no hallucinations, but the night was still disturbed. Frontal headache, dizziness, cramps in the limbs. The trembling persists and is more marked on the right side. Treatment; a jug of gentian tea; 30 grains of bromide of potassium.

March 13th. The patient is calm and tranquil and allows of a complete examination.

The trembling in both hands is continuous; whilst scarcely noticeable in the left it is very marked in the right, where it affects a partly regular rhythm, which has some analogy to the trembling of paralysis agitans. At times it is exaggerated, it ceases during sleep, but recommences if the patient although asleep moves in bed; this sometimes occurs under the influence of nightmares.

When stretched out, the right arm quickly becomes fatigued, and trembles much more; and the right leg when raised from the surface of the bed is tired much more speedily than the left, and also trembles more. In standing, the pressure is much less on the right than on the left foot; the patient supports himself with difficulty on the right foot, but jumps easily on the left.

Marked anæsthesia of the whole right side; the breath, which is scarcely felt on the right leg, is felt a little on the right arm; it is felt over the whole of the left side, even when slightly blown. The prickings of a pin are not felt so well as in the normal condition over the whole left side; on the right side, the hemianæsthesia is complete over the whole of that half of the body, face, neck, arm, chest, penis and leg. Tickling does not produce any reflex action, "it seems" says the patient "as if there were a piece of parchment between the skin and the finger, I scarcely feel it."

Temperature is ill appreciated on the right side, well perceived on the left.

The cutaneous sensibility to electricity is much weakened on the right side; retained on the left. The electric sensibility of the muscles is also diminished on the right side; the muscles contract on the right as well as on the left under the action of the induced current, but whilst on the right side one can pass a very strong current, so as almost to tetanize the muscles, without causing pain, on the left a weaker current excites, as in the normal condition, a sharp pain which prevents its strength being increased. The sense of touch is very much blunted, the patient passes the finger over a button of his uniform, or over a piece of money, without feeling the projections or distinguishing its form.

The mucous membranes likewise present an advanced degree of anæsthesia on the right half of the body. The right side of the uvula, velum palati and pharynx may be tickled without exciting the slightest nausea, or the least effort at vomiting; on the left side, on the other hand, reflex nausea occurs immediately. The feather of a pen may be introduced into the right nostril, or the external auditory meatus, without exciting the least tickling sensation; in the left, on the contrary, the sensation is disagreeable as in the normal condition. The conjunctiva, sclerotic, and even the cornea are but little sensible on the right side, one may pass a paper spill over the surface, without provoking the least movement of of the pupils, although, on reaching the cornea, the patient draws back his head a little, although he professes not to feel; and it is no doubt an instinctive movement, due to the perception by the retina of the foreign body. The general sensibility of the tongue is weakened on the right half, and taste is also much blunted on the right side;

saccharine and saline tastes are not appreciated; the taste of colocynth is barely perceived. On the left side, on the reverse, the substances excite the usual sensations. Odours are not distinguished by the right nostril, (orange flower water, essence of mint, camphor and vinegar, for example), whilst they give rise to the usual sensations in the left nostril.

Amblyopia on the right. With the ophthalmoscope the disc and retina appear normal, only the veins are slightly injected. Atropine dilates both pupils equally. There is no dyschromatopsy.

Hearing is impaired on the right side; the ticking of a watch is not perceived at a distance of 3 centimetres (1¼ inches) by the right ear, the left ear hears them at 80 centimetres (16 inches).

Heaviness in the head; pain, sometimes violent but not of shooting character, on the left side of the forehead; dizziness and sometimes vertigo. Dreams and nightmares during the night. Treatment. 30 grains of bromide of potassium; a jug of gentian tea.

The hemianæsthesia persisted for ten days without much alteration, then a slight improvement occurred, and on April 11th, sight appeared equal on the two sides; hearing remained slightly weakened on the right side; taste was still slightly blunted on the right, yet the flavours of salt and sugar were fairly well perceived, and smell appeared normal on both sides. The general sensibility appeared blunted on the right side. Trembling had ceased on both sides. The patient was quiet, occupied himself in the daytime, and slept peacefully.

The treatment at first consisted in the employment of iodide of potassium in the dose of 15 grains a day, and bitter infusion.

The improvement was maintained, and the patient went out on April 16th.

Complete hemianæsthesia is rare in man, but the loss of tactile sensibility, accompanied by diminution of special sensibility in one or several senses, is of still more exceptional occurrence, and on this account our case presents some interest. But together with the anæsthesia, there existed some paresis and a very marked rhythmical trembling, especially in the earlier days of the attack. It is very probable that this combination of symptoms depended on some cerebral lesion; and, in spite of the rapid improvement which took place, the existence of such a lesion appears to me to be none the less evinced by the cerebral disturbances which dated from two years before, the left frontal headache, attacks of dizziness and hardness of hearing, on which the patient insisted with the greatest precision on several occasions. In his case too, we see symptoms appearing after the excessive use of absinthe (such as biting of the tongue, and involuntary passage of urine), which suggest the existence of epileptic attacks.

CASE XXXI. *Chronic alcoholism with subacute attacks; first admission to St. Anne, on Sept. 15th, 1869; after abusing some policemen in the street.—Second admission, Feb. 15th, 1872; after insulting and threatening some soldiers.—Third admission, April 17th, 1873; after striking passers by, going into other people's houses, upsetting the furniture and pursuing an imaginary enemy.—Left hemiplegia; left hemianæsthesia of general sensibility and of the senses; equal aggravation of the hemiplegia and hemianæsthesia.*

D. S—— æt. 49, married, a sempstress, admitted into St. Anne, on Sept. 25th, 1869.—She is in the habit of drinking to excess, takes black currant wine every morning; has slept badly for some time, sees shadows on the wall, hears threats, becomes terrified and thinks she is pursued. Whilst in the street she insulted some policemen, saying that they wanted to assassinate her. The trembling of the hands, which was very marked on her arrival, improved on the following day. The head was heavy; at times attacks of dizziness, with numbness of the limbs occurred. The painful hallucinations pretty soon disappeared during the day, but sleep was disturbed and the patient retained some notions of persecution. By degrees D—— got better, and was in a fit state to leave the Asylum in March 1870.

On Feb. 15th 1872, she was admitted for the second time into St. Anne in an attack of alcoholic delirium. She had returned to her drunken habits; and on the day of her arrest, she had insulted and threatened some soldiers who were passing quietly by in the street, and who, she said, called her a Communist and wanted to kill her. For some days before, she had slept badly, had dreams and hallucinations, saw guns pointed at her, robbers, assassins, hideous heads on the wall which she tried to cover with clothes, but which reappeared directly at the side. She trembled so much in the hands that she had been obliged to give up the work of sewing; she had quite lost her appetite and sometimes vomited in the morning. At the end of some months she went out cured.

On April 17th 1873, she returned for the third time to St. Anne; for some days before this she had been selling ginger-bread at a fair, which was the occasion of redoubling her excesses. On April 15th, she lost her baskets and wares, and ran about the streets knocking against and striking the passers by; she went into the houses of people whom she did not know, went into a room, hunted under the bed and behind the furniture, upsetting every-thing, hearing, she said, a woman who was abusing her and who was hiding there.

On admission, she is very restless and terrified, hears the noise of firing of musketry, and of cannons, and voices jeering her and threatening to throw her into the water; she sees grimacing heads which change colour, multiply and alternately enlarge and diminish; she sees her own head assume colossal proportions, and then diminish and become quite little. She has considerable trembling of the hands; suffers from frontal headache, vertigo and noises in the ears; and brings up a very little phlegm in the morning. On auscultation of the heart a murmur is heard with the first sound.

After a few days, having become quiet, D—— allows herself to be examined more willingly, and we find an incomplete hemi-plegia with hemi-anæsthesia of the left side. Trembling is slight-ly more marked on the left side. The left leg is weakest and cannot be completely raised from the ground, whilst, on the con-trary, the patient easily hops on the right foot; the right hand carries the index of the dynamometer to the 22nd division, the left only to the 11th. On the left side, the breath, tickling, cold or moderately warm bodies are not felt; a pin passing through the skin is scarcely felt; on the right side, sensibility is normal in all its forms. The nasal, buccal and conjunctival mucous surfaces are likewise anæsthetic on the left side. The left ear hears the ticking of a watch at 10 centimetres, these are heard at 50 centi-metres by the right. The left eye can only read letters of 3 milli-metres, the right reads letters of 1 millimetre. Colours are more rapidly and readily distinguished by the right eye. Ophthalmoscopic examination shews nothing abnormal in either eye. The left nostril recognizes neither orange-flower water, camphor, essence of peppermint, nor musk; these substances are perfectly recognized by the right nostril. The left side of the tongue remains insensible to salt, sugar and colocynth, which are promptly recognized as soon as they are placed on the right side of the organ.

The paralysis of movement and sensation underwent slight increase in the following months. The patient could sew, but was obliged to watch the left hand incessantly, as she pricked herself as soon as she ceased to look at it, and did not notice it till she saw drops of blood fall on her work.

Nov. 13th, 1873. The hemiplegia and hemi-anæsthesia have made evident progress. The patient stands with difficulty on the left leg; in walking, it sometimes suddenly gives way; the right leg remains firm. The left hand moves the index of the dynamometer to the 8th division, the right to the 20th. Pricking with a pin, pinching, tickling, cold and hot objects are not felt on the left side, except in the hypochondrium, where tickling and pinching excite a slight sensation. Sensibility appears normal on the right side. The muscular sense is weakened on the left side, so also is the electro-muscular sensibility. A watch is now not heard by the left ear, unless it is applied directly to it and touches the lobe of the ear. On the right side it is heard at 70 centimetres (28 inches). When the right ear is stopped, a loud voice, and the sound of cutting of a saw are not heard by the left ear; hearing is normal with the right ear. The right eye speedily distinguishes all colours, the left recognizes them slowly and with greater difficulty; blue is taken for grey. The acuity of vision is less in the left eye, which cannot read letters smaller than 3 millimetres. Different solid substances placed on the left side of the tongue do not produce any sensation, whilst all savours are recognized on the right side. The left nostril does not smell anything, whilst the right discriminates perfectly orang-flower water, camphor, essence of mint, and musk.

A fortnight ago, a pin broke into D——s left hand, without her knowing anything about it. There are attacks of dizziness and vertigo, and flushings of the head, humming and whistling noises in the ears, most marked on the left side; sometimes a sensation of cold through the whole side. Intelligence is slightly weakened and the sleep is still disturbed from time to time by nightmare.

CASE XXXII. *Chronic alcoholism with subacute symptoms.—Left hemiplegia with hemi-anæsthesia of general sensibility and of the senses; improvement.—Persistent anosmia on the left side.*

Louis S—— æt. 46, a fruiterer, admitted to St. Anne on Feb. 28th, 1868, having been twice before under treatment at the Bicêtre, for symptoms of alcoholism. On his arrival, he presented the characters of an attack of subacute alcoholic delirium engrafted on chronic alcoholism. There was incomplete hemiplegia on the left side, weakness of the arm and leg, and lowering of the lip, but neither the general sensibility nor the special senses were examined.

Between 1868 and May 1873 he came into the asylum four times, for similar delirium. At the latter date it is noted that he had numerous and painful hallucinations of all the senses; he saw cats and rats; whilst at home he broke up some bottles to stop up holes which he saw in the wall; he hears the voices of his friends, bells and the firing of guns; he smells disagreeable odours; says that they put poison in his food which gives him colick; he feels prickings of the skin, numbness and cramps in the limbs.

The left side is the weaker, the right hand reaching the 34th division of the dynamometer, the other hand the 26th; he jumps less easily with the left leg. Throughout the whole of the left side, touching, pinching, pricking and temperature are less well felt; the anæsthesia is more marked on the arm than on the face and leg. The left nostril smells neither orange-flower water, camphor nor vinegar, which are perfectly recognized by the right nostril.

The left side of the tongue distinguishes neither sugar, salt, aloes, nor colocynth, though these substances are tasted very rapidly on the right.

The sight is slightly weakened on the left side. The left ear does not hear the ticking of a watch when further off than 20 centimetres, the right hears them at 60.

The hemiplegia and hemi-anæsthesia improve side by side, and at the end of two months, nothing definite is to be observed except very marked anosmia on the left side; the patient smelling neither camphor nor vinegar with the left nostril; the right nostril, on the other hand, distinguishing them very well.

ARTICLE VI.

Treatment of Chronic Alcoholism.

What we have already said with respect to predisposed patients, who are affected with alcoholic
delirium with frequent relapses, will allow us to
refrain from entering into any long detail in this
place. Confinement, or at any rate vigilant supervision, when, owing to improvement, or during a re-
mission the patients are allowed to live outside, appears to us to be indispensable. These patients, in
fact, as we have already had occasion to remark,
weakened in their intellectual, moral and affective
faculties, are yielded defenceless to the caprices of
their instinctive appetites. It is absolutely necessary
to prevent them from indulging in fresh excesses;
alcohol, in fact, exerts a most baneful influence on
their degenerated constitution, and weak doses of
the poison are sufficient to excite grave symptoms
in them, and even to endanger life. In the Asylum
itself, hygienic measures will play the greatest part in
the treatment; and for the chronic alcoholic, who has
not yet reached an advanced degree of dementia,
an easy occupation, requiring as far as possible some
little change of position and exercise in the open air,
should be chosen. But it will be necessary, if these
patients are engaged in agricultural labour, most carefully to avoid sending them into the fields, either
in summer under the rays of a burning sun, or in
winter at the time of the most intense cold. Great
heat and severe cold are equally dangerous for
these patients. It must not be forgotten, indeed,
that in chronic alcoholics the cerebral circulation
is greatly embarrassed, that the atheromatous state

of the large vessels, and the fatty degeneration of
the capillaries frequently give rise, even apart from
any external cause, to ischæmic or congestive disturbances—of which the attacks of giddiness, the
noises in the ears, and dizziness with heaviness and
weight of head, are the symptomatic expressions.
The same considerations are equally applicable to
chronic alcoholics who are progressing . towards
general paralysis, or who have already become
paralytic.　Diffuse sclerosis is well known to be
eminently favourable to the occurrence of disturbances of the circulation.

This morbid tendency demands certain precautions.　It will be necessary to keep a watch on the
habits, to regulate the ofttimes voracious appetite of
the patients, to exclude food which is difficult to
digest, and above all to avoid heavy meals in the
evening : so as to prevent indigestion, indolence and
over-fulness of the stomach, which are so favourable to congestive determinations of the blood to the
brain.

Led by similar considerations, recourse will be
had to revulsives to the skin and to the alimentary
canal; either by mustard foot-baths, or friction of
the lower limbs with croton oil or tartar emetic oint-
ment, which M. Baillarger often uses for analogous
indications in the early stages of general paralysis.
Aloes will be found very useful, not only on account
of its special tendency to act upon the pelvic organs,
but from its marked influence upon the course
of hæmorrhoidal congestion.　Saline and purgative enemas will also assist this treatment; we
may also, if need be, have recourse to the employment of hot cupping to the nape of the neck, and
also to dry cupping over the spine and lower limbs.

Tonics of all kinds, open air and hygiene, dieting, and therapeutic agents must furnish the basis of medical treatment. Bitter infusions, as we have already mentioned, given either between or at meals, are useful. Quinine wine after meals, and extract of quinine will also be beneficially employed here. The mixture of syrup of quinine, syrup of orange-peel, and Bordeaux wine, in the dose of 2 to 3 ounces after meals, has seemed to us of some value. Iodide of potassium, in the dose of 15 to 45 grains a day, will be especially useful in the cases of chronic alcoholism with a tendency to general paralysis. We are aware that Marcet* has recommended the employment of oxide of zinc, but it is especially in simple alcoholic delirium that the medicine appears to have been of use, that is to say, in cases where rest and good hygiene bring about a cure.

Magnus Huss has seen the epigastric pains and pyrosis diminish under the influence of the empyreumatic oil (*fermentoleum solani*), which gives to potato brandy its peculiar flavour, given in the dose of 1 to 2 minims, 5 or 6 times a day, either in a draught or in pills.

And lastly, M. Foville,† recalling the action of arsenic in intermittent fever, its action in diabetes, and in Basedow's disease, believes this drug to be useful in all cases where the symptoms may be ascribed to defective action of the great sympathetic, and he recommends it in the congestive form of general paralysis. We do not know

---

* Marcet. *On Chronic Alcoholic Intoxication.* London 1860.

† Foville (A.) Etude clinicale de la folie avec prédominance du délire des grandeurs. *Mém. de l'acad. de Medicine.* Paris 1870-71, p. 448.

whether the physiological action of arsenic is such as M. Foville would be inclined to attribute to it, but we know that this drug, in M. Baillarger's hands, has been useful in some cases of general paralysis in the first stage, where Pearson's solution was prescribed in the dose of 15 to 30 drops a day. In chronic alcoholism, tending to general paralysis, which is in many respects analogous to the congestive form of general paralysis, arsenic might therefore perhaps be employed with advantage.

CHAPTER V.

In the preceding chapters, we have endeavoured to answer the question set by the Academy: we have studied the various forms of alcoholic delirium, and have pointed out the treatment suitable for each form. In order to complete our work, it remains for us to compare alcoholic delirium as seen without any complication, with the numerous modifications which spirituous liquors produce in patients affected with various forms of insanity or certain diseases of the nervous centres.

We cannot here enter into all the developments of which this study would allow, and which are undoubtedly of very great clinical interest, but which are only indirectly related to our subject.

In a general way we may say that when in the presence of other affections of the nervous centres, alcohol acts in one of two ways, either as an excitant, in which case one sees the phenomena peculiar to the affection developed with unusual activity; or, exceeding the limits of an ordinary exciting cause, the alcohol acts as an actual poison and produces its own delirium: in the latter case the toxic effect is sometimes so energetic that its manifestation may completely mask the symptoms of the pre-existing disease, which in this case, do not shew themselves again until after the disappearance of the alcoholic symptoms. This fact is observed in the most remarkable manner in general paralysis; and if the paralytic is in the first stage of the disease, he has all the appearance of a simple alcoholic ; but when the hallucinations and the accompanying delirium have disappeared, when the trembling due to the poison has improved or is about to cease, at the same moment one sees the signs of general paralysis gradually reappear, one by one, as it were ; we shall return to this subject presently.

As we cannot take up all cases, we shall limit ourselves to some mental affections, and afterwards say a few words on the association of alcoholic symptoms with intercurrent diseases.

ARTICLE I.

Dipsomania.

All authors, now-a-days, distinguish dipsomania from alcoholism ; the former is a peculiar form of instinctive monomania, having its source most frequently in heredity ; alcoholism, on the contrary, is a simple poisoning which appears in the same manner in all persons, and we may add, in animals, as in man.

M. Trélat brings out clearly the differences which exist between the alcoholic and the dipsomaniac.* " Drunkards", says he, " are people who get drunk when they find an opportunity of drinking." "Dipsomaniacs are diseased persons who get drunk whenever their attack seizes them."

The attack of dipsomania, says Marcé, commences with sadness, moroseness, headache, and præcordial anxiety, and then the necessity of drinking makes itself powerfully and irresistibly felt. Dipsomaniacs feel the attack coming on ; and whilst deploring

* Trélat. *De la folie lucide*, 1861, p. 151.

their powerlessness to overcome this morbid crav-
ing, they do not leave off drinking until either
the attack is over, or the interference of their
neighbours puts an end to their excesses by an
enforced isolation.*

The dipsomaniac, before he drinks, is in a condition analogous to that of a melancholic; he is sad,
restless, sleeps badly, loses his appetite and has
præcordial anxiety; he is an ordinary insane person, but after a few days indulgence, intoxication
occurs and the dipsomaniac presents himself with the
alcoholic delirium which we know; he has painful
hallucinations, trembling, insomnia, gastric disturbance, &c.; and it is only after the disappearance
of the acute symptoms that the diagnosis becomes
complete.

The following case will give us an example of
this.

CASE XXXIII. *Attacks of dipsomania for 23 years, separated at
first by intervals of eighteen months, now by intervals of only four
months.—Attempted suicide.—Attack of alcoholic delirium.*

The patient D——, wife of V——, aged 57 years, a shirt
maker, came into St. Anne's on October 10th, 1869, with the
following medical certificate : " Alcoholism, alarming hallucinations; anguish, fright, insomnia."

On admission D——, is excited, talking, crying out and frightened; she hears assassins who want to strike her; sees by her
side the heads of the victims of Pantin; thinks she is covered
with vermin and shakes her clothes; hears the voices of her
relations; people are knocking near her, &c. The hands are
tremulous; tongue white; pain in the epigastric region. Passed
a sleepless night, the hallucinations being incessant.

At the end of five days the delirium has disappeared, the
patient is sad and depressed, but quiet, and employs herself in the
day-time; at night the sleep is still bad and disturbed with nightmare.

* Marcé, *loc. cit.*, p. 388.

The information with respect to her is as follows; for more than thirty years, at certain periods, the patient has become sad, not taking an interest in anything, feels weak, and incapable of any work, sleeps badly, has no appetite, and suffers from a sense of discomfort in the stomach which is increased by the sight of food; she has burning thirst, and begins to drink wine on the first day. Next day she gets a bottle at home, hides it as she goes down stairs, goes and gets it filled at the wine merchant's, runs quickly home and shuts herself up, and drinks until she drops down. As soon as the drunkenness begins to go off, she hastens to drink again, and so goes on for several days.

After the attack, she reproaches herself, and is horrified at herself, and returns to her regular way of living and sober habits. These attacks were at first separated by intervals of from fifteen to eighteen months; at the present time they occur nearer to each other, leaving an interval of only three to four months between them. Twenty years ago, the woman tried to drown herself at the onset of one of these attacks of dipsomania.

During a long period, at the time when the attacks of dipsomania were separated by intervals of eighteen months, drunkenness alone accompanied the drinking attacks ; but later, when they came nearer together, and the alcohol could act in a more continuous manner, hallucinations and delirium developed themselves in their turn.

The following case, borrowed from M. Trélat's work, brings out still more clearly the difference which exists between the dipsomaniac and the alcoholic.

CASE XXXIV.* *Dipsomania. Mother and uncle dipsomaniacs.*

Madame N——, was a person of serious character. During her life she had several establishments which always failed from the same cause. Though habitually regular and economical, she was seized from time to time with irresistible attacks of drunken monomania which made her forget everything, duties, interests, and family ; and ended by plunging her from a condition of great comfort into complete ruin.

It was impossible to listen to her account of the efforts which

* Trelat, *loc. cit.*, p. 160

sho had made to cure herself of a tendency which had always been so deadly in its effects upon her, without being moved with lively compassion for her. When she felt her attacks coming on, she put into the wine which she drank the substances most calculated to inspire her with disgust for it. But in vain. She even mixed excrement with it. At the same time she abused herself, saying, 'Drink then, wretch, drink then, sot, drink, villainous woman, who forgettest thy first duties and disgracest thy family.' Her passion, or rather her disease was always stronger than her self-reproaches, and than the disgust with which she tried to inspire herself. During the later years of her life she was successfully operated upon for strangulated hernia, and died at last of disease of the heart.

ARTICLE II.

General Paralysis.

General paralysis is one of the diseases which are most frequently associated with alcoholism. General paralytics are not only induced to drink by the peculiar characters of their delirium, but they are led also themselves to seek for opportunities of drinking; and we may also observe in them all the degrees of acute alcoholic intoxication. Sometimes when alcohol is taken in a small quantity, it excites or stimulates that morbid activity which is already so prompt to display itself in paralytics; at other times, exceeding this limit, it provokes illusions or some hallucinations in addition to the excitement; and, finally, in a still higher degree, it gives rise to an attack of alcoholic delirium which masks both the physical and intellectual symptoms of the general paralysis, rendering all exact appreciation of the mental condition impossible; whilst by its hallucinations it excites delirious ideas which appear most prominently, and hide all the other mental

* Trélat. *Loc. cit.* p. 160.

R

phenomena. On the other hand, the alcoholic trembling being much the more marked overbears the motor phenomena of the paralysis, and in some cases when the tremors involve the muscles of the face, even the embarrassment of speech loses its peculiar character.

The three following observations furnish us with examples of the kind.

CASE XXXV. *Attack of alcoholic delirium; at the end of three days, diminution of the alcoholic symptoms, and appearance of symptoms of general paralysis.*

Constant V——, æt. 40, a labourer, came into St. Anne on October 10th 1869, with the following medical certificate : "Alcolism, terrifying hallucinations, fear of assassination, tendency to violence." V—— appears in a state of violent excitement, goes about everywhere ; tries to seize imaginary creatures ; hears voices threatening him, sees armed men who are trying to injure him ; perceives animals, rats ; cries out, quarrels, and is abusive.

Considerable trembling of the arms and legs ; tremors of the muscles of the face ; pupils contracted ; skin bathed in perspiration.

The hallucinations disappear at the end of five days, but some ideas of persecution remain, which remind one of those of alcoholic delirium ; he fancies that some people are trying to do him an injury, that he has been knocked down and stunned. But at the same time, some ideas of self-satisfaction show themselves, he earns, he says, a great deal of money, there is no one else like him in the universe, he belongs to a good family, they are all well known ; he is vigorous, intelligent, &c.

The trembling of the hands has diminished, the muscles of the face are at rest, but when the patient speaks we can distinguish perfectly the peculiar hesitation of the paralytic.

Moreover, the history of the case informs us that V—— has not worked for the last five or six months, that his memory and and his mental faculties have become impaired, and that he has had ambitious ideas for the last two months ; that for a long time he has complained of pain in the head and giddiness, and of numbness of the limbs, and that his speech has become embarrassed on several occasions since about four months ago.

The symptoms of general paralysis became gradually more and more marked.

This patient retained for several days some ideas of persecution, and sometimes in the night he also heard voices of people who wanted to injure him.

Another paralytic, who came in with very marked alcoholic symptoms which masked the symptoms of general paralysis, retained, so long as a year after admission, some hallucinations at night and some ideas of persecution; he heard men come into his cell to steal his wine from him; he saw his servant stealing things from his cupboard; and he thought that some one wanted to seize upon his fortune, and paid the authorities to prolong his confinement. At the same time his general paralysis was already in an advanced state, and the embarrassment of speech was enormous.

CASE XXXVI. *Attack of alcoholic delirium; disappearance of the toxic symptoms at the end of six days.—Signs of general paralysis.*

The patient, Henry C——, aged 49, a moulder, was admitted into St. Anne on October 9th, 1869, with the following medical certificate: "Delirium tremens; attack of frenzy." On his arrival, C—— is frightened, hides himself in corners, thinks that people shoot at him; a band of assassins is at his heels; he sees men climbing up to the window by ladders; he is pursued; sees cats running about the room, &c.

Trembling of the whole body; tremors of the muscles of the face; jerking style of speech.

At the end of two days the hallucinations have disappeared, the trembling has almost completely ceased, and one finds a weakening of the faculties and of the memory, with ideas of satisfaction: he says he is a very good workman; he will earn good wages; he will take the first place; he fears nobody; he is very strong; smiling, contentment.

Hesitation of speech, left pupil the larger.

The symptoms of general paralysis undergo a progressive increase.

His sister had told us, moreover, that for fifteen months his character had changed, he got in a passion without any reason; at other times he let himself be treated like a child, that he lost his memory and did not do his work well any longer; from the first, his wages had been reduced, and for three months he had been unable to work at all. He often suffered with his head, and had pains in the arms and legs.

In the following case we see alcohol acting in the character of an excitant.

CASE XXXVII. *Symptoms of general paralysis with excitement.— Trembling of the hands, hallucinations at night.*

Louis H——, æt. 40, a wine merchant, came into the asylum on March 25th, 1869. On admission, he shakes hands with everyone, he is loquacious, talks, vociferates and is incoherent; says he wants to make everyone happy, that he will not be a wine merchant any longer, he is going to become inspector-general of police to the Emperor; he will build; he has house-property. He goes about hither and thither, busies himself with everything, and is incessantly in motion. Considerable trembling of the hands, hesitation of speech, right pupil larger than left.

During the night he hears people trying to get into his room; he is busy with his customers, talks to them and has them served.

The excitement ceases at the end of a few days; the trembling of the hands almost entirely disappears; speech is still embarassed; pupils unequal. The signs of general paralysis become gradually more marked.

One might have thought at the time when the patient was admitted, that he was suffering from one of those attacks of excitement which so frequently occur in the course of general paralysis, but there was considerable trembling of the hands, which revealed another cause; the hallucinations at night also testified in some measure to the influence of the poison. Moreover, the history told us that although H—— had been ill for six months, he remained in his shop, and drank a good deal.

For four months he had had ambitious delirium and hesitation in speech. The impairment of intelligence dated as far back as a year, during which period, headache, numbness of the hands, and pains in the legs had been observed.

Under some circumstances the excesses in drinking act solely as excitants, without giving rise either to trembling or hallucinations, but impressing a character of greater activity on the delirium, and arousing a sense of need of movement, and agitation, such as one sees under the influence of various kinds of stimulation.*

ARTICLE III.

Delusions of Persecution.

When alcoholism is associated with insanity with ideas of persecution, we can in the majority of cases distinguish between the preexisting delirium and that due to the intoxication.

M. Laségue, recalling in reference to this subject, the character of the insanity of persecution, adds :— " Not only is the field of the disturbances thus limited, but within this sphere nothing is changeable. The monotonous delusion tends to become stereotyped. The conspiracy once organized is so well organized that it plays its part almost without variation, or at any rate without any great change.

* The combinations of general paralysis with alcoholism may sometimes lead us into error, especially as alcoholism, as we have seen, plays an important part in the etiology of general paralysis. M. Lunier, in his *Recherches sur la paralysie générale progressive* (Researches on progressive general paralysis) 1849, p. 15, insists with some reason on this particular point, and after shewing clearly the difficulty of diagnosis, he reminds us that general paralysis may have been mistaken for a chronic delirium tremens.

The man who is accused of being a thief is always accused of having stolen, he who is magnetized is always more or less magnetized, and he who is being poisoned by means of his food continues to be the subject of poisoning of a similar nature."*

In alcoholic delirium, as we know, on the contrary, it is the changeableness, multiplicity, and variety which characterize the delirious conceptions.

The following case will give us an idea of this combination.

CASE XXXVIII. *Insanity of persecution, presenting two periods of excitement, during which hallucinations resembling those of alcoholic origin appear.*

C. R——, a married woman, æt. 33, admitted to St. Anne on Feb. 16th. This patient, the daughter of an unmarried woman and not knowing her father, imagines herself on several occasions to be the offspring either of rich proprietors in her neighbourhood, or of officials occupying high positions; other delusions are subsequently added to the first, she believes that her origin is hidden from her in order to deprive her of a fortune, that people organize conspiracies against her, and that documents have been made away with, &c.

For some time past she has been drinking, and takes aniseed liqueur in rather large quantities. The nights are bad, the delusions of persecution become more active: *they accuse her of having stolen; men and women run after her, abuse her and want to send her to prison; they threaten her; she sees phantoms and ghosts in her room.* Terrified and wanting to make an end of it, she throws herself from the window.

She arrives at St. Anne with an enormous contusion of the left buttock; she is restless and frightened; people accuse her, she says, and want to injure her; they are tormenting her.

During the night the hallucinations are more frequent, she perceives shadows passing over her bed, she sees faces at the door, and behind the window-panes; phantoms prowl around her bed, &c.

The trembling of the hands is very slightly marked.

The hallucinations and agitation persisted for a fortnight, but

* Lasègue, *loc. cit., Arch. génér. de Méd.,* June 1869.

they were followed by insanity with delusions of persecution, which had not entirely disappeared when the patient went out of the hospital four months later.

After returning to her husband, she occupied herself pretty regularly with household affairs, ate with an appetite, and slept tranquilly; she continued, however, to talk of her birth and of people who wanted to rob her, and she said that she knew now the residence of her father. On several occasions, again and again, she went to the house of a public functionary, a director in a public office, called him her father, and exhibited the warmest affection for him. Seeing that she was mad, they took her in charge two or three times, treating her with all consideration, and then very gently dismissed her.

Still later, when the door was closed, she got into a violent passion, abused the porters and created a disturbance.

She soon began to drink again: the delirium became more active, she no longer stayed in-doors, but, dressed out in her best clothes, she roamed about the streets, looking for and expecting to meet the person whom she called her father; several times she got much excited at home; she slept badly, heard cannon, and the firing of musketry, saw black and white birds fluttering about in her room; saw the devil; got up and opened the windows, shouted, was much excited, &c.

She again entered the asylum for the second time, on April 7th, fourteen months after her first confinement.

She is calm enough in the daytime, but at night she sees shadows and phantoms, and hears voices threatening her; everything is moving about under her room, they are knocking and making a dreadful noise, it is like hell, she says. The hands shew a scarcely appreciable trembling.

At the end of a fortnight, the painful hallucinations at night had ceased, the ideas of persecution remaining as they had been previous to this period of excitement.

In this patient we find a persistent delirium, always the same in character and tending to become systematized; in two relapses it becomes exaggerated under the influence of alcoholic stimulation, then it becomes combined with mobile, multiform and variable hallucinations, which have come on as the result of excesses, and disappear with the cessation of the latter. These are two sets of symptoms of different origin, which may, moreover, be distinguished from each other by analysis.

ARTICLE IV.

Imbecility.—Epilepsy.—Senile Dementia.

Imbecility.—Alcoholism is frequently combined with imbecility, and we may see every degree of intoxication in the imbecile. In these originally degraded subjects, the action of alcoholic liquors is much more powerful. Their drunkenness is frequently accompanied by attacks of violent passion and sometimes of fury; and whilst dementia, on the one hand, comes on more rapidly in them as the result of excesses, on the other hand, of the small number of imbeciles whom we have seen become paralytics, the majority were addicted to drinking.

Epilepsy.—Epileptics, even more than imbeciles, are very prejudicially affected by alcoholism, and if we find that there are exceptional cases where the abuse of drink does not increase the frequency of the attacks, the majority of patients, on the contrary, are especially subject to them when they have been drinking, and we have seen men who counted their attacks by the number of 'sprees,' as they call them, which they have had.

The epileptic alcoholic is one of the most dangerous of patients; he adds to the impulses, sometimes so terrible, to which he is subject from his disease, those which he draws from his intoxication. We need not say that we mean here to speak only of simple epilepsy, usually of hereditary origin, and as a rule shewing itself in the patients before the age at which drinking excesses begin. What we have already said on absinthic epilepsy, and on the epileptiform attacks which supervene in chronic alcoholics, will permit us to refrain from entering into further details ; the difference between these forms of convulsion having been sufficiently demonstrated.

Senile Dementia.—In senile dementia alcohol may act as an excitant, or may give rise to its peculiar delirium when the dose is sufficient. The combination of alcoholism with senile dementia, although less frequent than the association of this intoxication with general paralysis, is, however, not rare ; and we have several times seen patients affected with senile dementia, arrive at the asylum in an attack of alcoholic delirium.

ARTICLE V.

Alcoholic delirium developed under the influence of intercurrent diseases.

The drunkard who can, if not with impunity, at any rate without any very obvious appearance of symptoms, give himself up to habitual excesses in drinking, ceases to enjoy that immunity, as soon as ever the physiological equilibrium is destroyed by the blow of an intercurrent disease. The vital powers, which previously sufficed to resist the poison, being diminished, the support gives way, and the symptoms are produced. Nor is it only the depressing effect of disease which can bring about this result, for any cause of debility may give rise to it, such as insufficient food, fatigue, and excesses of every kind. Nor must we lose sight of the fact that these accidents will develope with so much the greater facility, in proportion as the poison has acted for a longer time, and the system consequently has become more profoundly altered.

We will recall the fact that subacute symptoms were developed during convalescence, in some of our patients, under the influence of a slight attack of ill health; and in L——, (Case xii), in particular, the hallucinations recurred, accompanied by delirium and nightmare, a long time after they had disappeared. All the symptoms ceased, moreover, at the same time as the intercurrent febrile affection.

During the cholera epidemic of 1865, M. Mesnet had the opportunity of observing three cases of alcoholic delirium developed under the influence of that disease. " The period of reaction, says that author, even when free from all serious complications, sometimes presented unusual phenomena, the cause of which did not escape us. These were persons, who were surprised, as it were, by the epidemic in the midst of an as yet but slightly marked degree of alcoholic intoxication, and who would not have had any cerebral disturbance, if their health had not become suddenly affected; but, when the cholera had touched them, when their strength was exhausted by the vomiting, diarrhœa, and by their diet, the alcohol displayed itself in their exhausted system by disorders which are not without analogy in other affections."

" In three of them, we were able to observe, almost from the time of admission, a combination of symptoms almost simulating an inflammatory condition. One of them, who had been taken with vomiting and diarrhœa five days before, came to us on the sixth day with the skin warm, profusely perspiring, pulse 108, countenance animated &c., there were no symptoms of cholera except attacks of vomiting, diarrhœa and cramps, he never had either the coldness or the cyanosis. Then, about the second day, his intelligence, which had till then remained clear, was in turn affected; there was none of the slowness of ideas of cholera patients in the algide stage, there were, on the contrary, loquacity, incoherence in the ideas, and then some hallucinations of sight, which became more powerful in the evening, were exaggerated during the night, and tended to disappear in the morning. Under their influence the patient got up and tried to go out of the ward, not knowing either what he was doing or where he was; his hands were tremulous, his walk undecided, his look uncertain and stupified. For three days he remained in this condition; after which abundant perspirations appeared as a sort of crisis, and he entered into a complete convalescence."*

We might furnish several examples of alcoholic delirium (delirium tremens) which has come on under the influence of an intercurrent disease, such as erysipelas, pericarditis, rheumatism, pneumonia, &c.†

* Mesnet, Choléra de 1865; Saint Antoine Hospital (Male wards). *Arch. gén. de Méd.*, Feb. and March, 1866, p. 26.

† A. Voisin; *De l'état mental dans l'alcoolisme aigu et chronique; Ann. Médico-psych.*, 4ᵉ série; vol. iv, p. 55. In this work will be found mentioned two cases of articular rheumatism and a case of pneumonia with alcoholic delirium, observed under the care of Prof. Bouillaud at the Charité. (It may be questioned whether in these and other cases of the kind, the delirium is really always due to alcohol, although the symptoms and character of the delirium may closely resemble those of delirium tremens, especially in acute pneumonia, and rheumatic pericarditis and pneumonia. Further details and references to authorities on the subject, are given in a paper on "Insanity as a sequel of acute disease and hæmorrhage" in *St. Thomas's Hospital Reports*, vol. iv. 1873. *Trans.*)

But these facts would in reality possess only the merit of number, and we shall therefore confine ourselves to giving the record of one of the most common cases.

CASE XXXIX. *Habits of drunkenness.—Absence of previous delirium.—Pneumonia.—Attack of alcoholic delirium on the third day of pneumonia. —Tonic treatment.—Cure.*

Joseph T——, æt 38 years, a labourer, addicted to drinking, and subject to vomiting of phlegm in the morning, and trembling of the hands, which was sometimes very marked, was seized with shivering, a stitch in the side, oppression and fever, and was admitted to the Necker hospital on Oct. 8th, 1870. When he was brought to the hospital his wife stated that his reason was perfectly clear, that before leaving home he had occupied himself as well as usual with some little household affairs, and that nothing had led to the supposition of mental derangement.

On Oct. 11th, his wife went to see him at the hospital and found him in the most complete delirium ; moving about his arms, seeking imaginary creatures round his bed, and insects on the sheets; speaking to his companions and his friends, and busying himself about his work, &c.

Thus excited, as the patient disturbed the quiet of the ward, he was sent to St. Anne on Oct. 13th, 1870.

On his arrival he is talkative, frightened, cries out, at times he threatens, shakes his bedclothes, hunts and follows with his hand, on the wall and on the table, animals or insects which escape from him. He sees flames, and the scaffold giving way; people insult him and make false reports about him.

Trembling of the hands, muscular tremors in the arms and legs. Skin dry and warm on the face, slightly moist on the chest and abdomen ; tongue red at the tip ; embarrassed respiration ; viscid and rusty sputa ; dulness, bronchial breathing and fine subcrepitant râles at the lower angle of the left scapula; pulse compressible 105, T. 39·4° C (103° F.). Treatment: milk, beef-tea ; sweetened wine and water; gum julep, draught of 120 grammes ; liquid extract of quinine, 4 grammes; syrup of tolu, 30 grammes; three times daily. A blister had been applied before leaving the Necker hospital.

Oct. 14th. The night has been passed without sleep, the patient got up several times ; (he was covered with a cloth dress, see p.

149) which protected him from the cold; he held imaginary conversations with his wife and his comrades and talked of his workyard; on several occasions he saw shadowy phantoms and flames around him.

The trembling of the hands persists, but the muscular tremor is less marked. Abundant perspirations. Expectoration thicker, still rusty; bronchial breathing and fine subcrepitant râles; pulse large, compressible, 102 per minute. T. in rectum 38·6° C (101·9° F.). Treatment: Milk, beef-tea, sweetened wine and water. Prescription as before.

Oct. 15th. Three hours sleep during the night; delirium as before. In the morning the agitation is diminished; the patient still hears some insults, answers and defends himself. At times, he answers his wife and companions; he sees birds flitting over his head. Slight trembling of the hands. Abundant perspiration; respiration easier; expectoration thick, brownish in places; subcrepitant râles; pulse full, 108. T. R. 38·8° C (102° F.). Treatment: milk, beef-tea, wine and water; coffee, 4 ounces; 2 eggs; broth.

16th. Five hours of sleep altogether during the night. Hallucinations persistent, but less agitation. At the visit he answers questions addressed to him, but his attention is distracted by the voices of his friends or of people who he thinks are insulting him. Slight trembling of the hands. Sputa thick and yellowish; subcrepitant râles; pulse 96. T. 38·4° C (101° F.). Treatment: milk, beef-tea; wine and water; coffee; a cutlet and soup.

18th. Slept nearly all night; some hallucinations. The day is quiet, he begins to doubt the reality of his delusions. Sputa thick, yellowish. Pulse 92; T. R. 37·8° C (100° F.). Treatment: milk, wine and water, coffee; roast meat, morning and evening; soup.

The improvement continued on the following days; the nights were only disturbed by nightmare, and from Oct. 25th, T—— had no more delirious ideas. He went out *cured* on Nov. 13th, 1870.

Before the pneumonia T—— had no intellectual disturbance; his wife, when questioned on this point, was very positive about it. It was not till the third day of the pulmonary congestion that the symptoms shewed themselves, with all the characters of alcoholic delirium. When in his ordinary state of

health, although he was under the influence of a
long continued intoxication, T—— enjoyed sufficient
strength to enable him to resist the action of the
poison.   But the physiological equilibrium once
broken, the toxic agent, whose influence, though
latent, was none the less real, no longer encoun-
tering any resistance in the weakened system,
shewed itself ' by the usual disturbances.   It
was to maintain the strength and to restore the
regular functions of the whole system, that we had
recourse from the outset to tonics, such as quinine,
and also to milk, beef-tea and wine in small doses,
which possess the advantage of sustaining and gradually restoring the vital forces, without exciting
them.

INDEX.

A.

Absinthe
  action on animals, 22.
       man, 77.
  injection of essence, in small
    quantity, 23.
  injection of essence, in large
    quantity, 24.
  injection of essence, experiment,
    26.
Absinthic epilepsy, 25, 74.
Acute alcoholism, see Alcoholism.
Alcohol
  analysis for, 116.
  presence in tissues, 107.
Alcoholic delirium
  in man, 33.
  classification of its forms, 45.
  treatment, 93.
Alcoholism
  acute, 33.
  prophylaxis, 99.
  treatment, 92.
  chronic, 154.
  form tending to general para-
    lysis in, 176.
  hemianæsthetic form, 198.
  treatment of, 233.
Amaurosis, 42.
Ambitious delusions, 62.
Amblyopia
  in alcoholic delirium, 42.
Anæsthesia
  in chronic alcoholism, 199.
  acute alcoholism, 45.
  sensory, 43, 203.
Analysis
  for alcohol, 116.
Animals
  action of alcohol on, see Alcohol,
    absinthe, see Absinthe.
Anosmia, 207.
ANSTIE
  epileptiform seizures from injec-
    tion of alcohol, 77.
Aorta
  atheroma of, 195.
Arsenic
  use in chronic alcoholism, 235.
Asylum
  confinement in, in acute alco-
    holism, 94.
Atheroma of vessels, 195.

B.

BAILLARGER
  on meningeal hæmorrhage, 16.
  on time of occurrence of hallu-
    cinations, 39.

Benzoic chloride
  as reagent for alcohol, 116.
BERTHELOT
  method of analysis for alcohol,
    116.
BOUCHEREAU and MAGNAN
  statistics of epilepsy due to ab-
    sinthe, 76.
  statistics of tendency to suicide
    from alcoholism, 61.
BRIERRE DE BOISMONT
  on suicidal mania, 61.

C.

Camisole
  dangers of, 146.
  case illustrative of, 147.
CASPER
  on causes of suicide, 61.
Cerebral disturbance
  in chronic alcoholism, 158.
CHALLAND
  experimental injection of ab-
    sinthe, 24.
CHAUFFARD
  means of combating tendency to
    drunkenness, 101.
Chloral hydrate
  in febrile delirium tremens, 153.
Cholera
  alcoholic delirium during, 250.
Chromatoscopy
  retinal, 42.
Chronic alcoholism, 155.
  with tendency to general para-
    lysis, 176.
  hemianæsthetic form, 198.
  treatment, 233.
Colour
  defective perception of, 42.
Convulsions
  from absinthe, clonic, 25.
           tonic, 24.
Convulsive drunkenness, 6.
Cord, spinal
  condition of, 21, 114, 178, 180,
    &c.

D.

Delirium, alcoholic
  in man, 33.
  expansive form, 62.
  treatment of, 93.
  in dog, 8.

Delirium tremens
 simple, 39, 48.
 febrile, symptoms, 102.
 prognosis, 104.
 diagnosis, 141.
 treatment, 143.
Delusions of persecution, 245.
Dementia
 following chronic alcoholism, 160.
 senile, effects of alcohol in, 249.
Depression
 alcoholic delirium characterized by, 61.
Digitalis
 in febrile delirium tremens, 153.
Dipsomania, 238.
Diseases, intercurrent
 alcoholic delirium in, 249.
Dog
 alcoholic delirium in, 8.
 alcohol poisoning in, 18.
 drunkenness in, 7.
 delirium tremens in, 19.
Drunkenness in dog, 7.
 treatment of in man, 92.
Dyschromatopsy, 42.

E.

Electro-muscular sensibility
 diminution in chronic alcoholism, 202.
Elevation of temperature
 after epileptiform seizures, 198.
 in febrile delirium tremens, 194.
Epilepsy, absinthic
 experimental, 25.
 in man, 74.
Epileptiform seizures
 in chronic alcoholism, 159.
ERHARD
 on cause of auditory hallucinations, 41.
Erysipelas
 associated with alcoholic delirium, 251.
Exaltation
 delusions of, 62, 63.

F.

FALRET
 on general paralysis, 154.
False membranes
 in arachnoid, 16.
Febrile delirium tremens, 102.
 cases, 109, 118, 122, 127, 130.
 diagnosis, 141.
 prognosis, 104.
 treatment, 143.
Fever
 in delirium tremens, 104.
Fevers
 alcoholism associated with, 249.

G.

GALEZOWSKI
 on retinal chromatoscopy, 42.
GAMBUS
 on chronic alcoholism terminating in general paralysis, 163.
General paralysis
 chronic alcoholism ending in, 176.
 elevation of temperature in, 199.
General paralytics
 action of alcohol on, 241.
GUBLER
 on the properties of absinthe, 22
 on the treatment of delirium tremens, 143.

H.

HALLOPEAU
 on chronic diffuse myelitis, 157.
Hallucinations
 in dog from injection of absinthe, 26.
 alcoholic in man
 changeable character, 34.
 nature of, 35.
Hearing, defective
 in chronic alcoholism, 206.
Hemianæsthesia, 134.
 centre affected in, 210.
 in hysteria, 208.
Hemianæsthetic form of chronic alcoholism, 198.
Hereditary predisposition
 influence in alcoholism, 63.
Homicide
 tendency to, 61.
HUSS
 on classification of chronic alcoholism, 160.
 on treatment of do., 235.

I.

Imbecile,
 effects of drinking on, 248.
Incoherence
 in alcoholic intoxication, 6.
Intellectual symptoms
 in chronic alcoholism, 155.
Intercurrent diseases,
 alcoholic delirium in, 249.
Iodide of potassium
 in chronic alcoholism, 235.

J.

JOLLY
 on the prophylaxis of alcoholism, 99.

K.

KREMIANSKY
 on pachymeningitis, 15.

L.

LACAROSSE
  case of alcoholic delirium. 78.
LALLEMAND, PERRIN and DUROY
  on presence of alcohol in viscera,
  11, 108.
LANCERHAUX
  effects of chronic alcohol poisoning, 155.
LASÈGUE
  on hallucinations in alcoholism,
  35.
  subacute alcoholism, 95.
Liver
  condition in alcoholic poisoning,
  11, 12, 107.
LOLLIOT
  on chronic alcoholism as a cause
  of general paralysis, 183.
LUNIER
  on relations of general paralysis
  and delirium tremens, 245.

M.

Maillot, 149.
MARCE
  on action of essence of absinthe,
  22.
MARCEL
  on mental disturbance due to
  alcohol, 33.
MARCET
  on oxide of zinc in alcoholism,
  235.
Meningeal hæmorrhage, 125.
MESNET
  on alcoholic delirium in cholera,
  250.
MEYNIER
  on properties of absinthe, 22.
MIERZEJEWSKY
  on cerebral lesions in general
  paralysis, 194.
MOREAU
  on predisposition to delirium,
  63.
MOTET
  on effects of absinthe drinking,
  29.
  case of absinthic epilepsy, 79.
  on confinement in alcoholism,
  96.
Mucous membranes
  anæsthesia of, 204.
Muscular tremor in febrile delirium
  tremens, 105.
  weakness in do., 107.

N.

NEUMANN
  on Pachymeningitis, 15.
Nightmares
  from alcoholism, 40.

O.

Opium
  in febrile delirium tremens, 153.

P.

Pachymeningitis, 15.
Paraplegia, alcoholic
  in dog, 8, 9.
PERCY
  on convulsive drunkenness, 6.
Persecution
  delusions of, 245.
Pneumonia
  case of delirium tremens in, 249.
  from intoxication and exposure,
  20.
Pulse
  in febrile delirium tremens, 105.
  in general paralysis, 193.
PUPIER
  on action of alcohol on the liver,
  12.

Q.

Quinine
  in chronic alcoholism, 235.

R.

RUGE
  condition of kidneys in alcoholism, 14.

S.

Sclerosis of spinal cord, 157, 178,
  &c.
Senile dementia
  effect of alcohol in, 249.
Sensation, common
  derangement of, 44.
Sensory disturbances in alcoholism,
  39.
Sight,
  affections of, 205.
Smell,
  derangement of sense of, 43, 207.
SPERLING
  on hæmorrhagic pachymeningitis, 16.
Spinal cord, *see* Cord.
Straitwaistcoat
  dangers of, 145.
Suicide
  drunkenness as cause of, 61.
Suicidal tendency, 61.

T.

TARDIEU
  on meningeal hæmorrhage, 11.
  phosphorus poisoning, 166.
Taste
  derangement of sense of, 43, 207.

Temperature
  in febrile delirium tremens, 104.
  lowering of by alcoholic poison-
    ing, 10.
Tissues
  presence of alcohol in, 107.
Treatment
  acute alcoholic delirium, 93.
  chronic alcoholism, 233.
  drunkenness, 92.
  febrile delirium tremens, 143.
  prophylactic, 99.
Trembling, characters
  in febrile delirium tremens, 105.
TROUSSEAU
  on action of absinthe, 22.

V.

Vertigo from absinthe, 25.
Vision
  derangements of, 41, 205.
VULPIAN
  on false membranes surrounding
    hæmorrhages, 17.
  seat of sensory perceptions, 210.

Z.

Zinc oxide
  in simple alcoholic delirium, 235.